**Ben Stacy Jerrik (Ed.)**

**SendStation Systems**

Ben Stacy Jerrik (Ed.)

# SendStation Systems

## Macintosh, iPod

Part Press

**Imprint**



Publisher:
Part Press is a trademark of
International Book Market Service Ltd., 17 Rue Meldrum, Beau Bassin, 1713-01 Mauritius
Email: info@bookmarketservice.com
Website: www.bookmarketservice.com

Published in 2012

Printed in: U.S.A., U.K., Germany. This book was not produced in Mauritius.

**ISBN: 978-613-8-88630-3**

# Contents

## Articles



## References

# SendStation_Systems

| | |
|---|---|
| **Type** | Private |
| **Industry** | Computer hardware |
| **Founded** | 1997 |
| **Headquarters** | Frankfurt/Main, Germany |
| **Key people** | André Klein, Founder, President and Head of Development |
| **Products** | Computer hardware |
| **Revenue** | unknown |
| **Employees** | unknown |
| **Website** | www.sendstation.com [1] |

**SendStation Systems** is a manufacturer of computer and iPod accessories. The company was founded in 1997 in Frankfurt/Main, Germany by current President André Klein.

The name "SendStation" has its roots in a Macintosh-based turn-key video fax system the company created and sold between 1997 and 2000, and is a word play made up from *to send*, *station* and *sensation*. Back then (several years before public broadband internet access became available and affordable) *the* SendStation was the first system of its kind which allowed filmmakers, advertising agencies and industry clients to easily transfer full-screen-full-motion review copies of TV commercials within minutes around the globe, rather than using overnight couriers. Customers included companies like Audi, Wrigley and BBDO.

## From system developer to accessory manufacturer: PocketDock

In 2003 SendStation entered the hardware market, and became one of the initial five, officially authorized accessory manufacturers for the Apple iPod. Its first hardware product, the PocketDock FireWire adapter, allowed iPod owners to connect the device via the then freshly introduced 30-pin iPod dock connector to a standard 6-pin FireWire cable for sync & charge. It was one of the best-selling add-ons on the young market for iPod accessories.

As USB 2.0 became standard and iPods with video capabilities emerged, additional models followed. In order of release (most recent on top):

- PocketDock Line Out Mini USB
- PocketDock AV (USB, Audio Line Out, Composite & S-Video)
- PocketDock Line Out USB
- PocketDock Combo (USB+FW)

- PocketDock Line Out FW
- PocketDock FW

Other products include iPod car chargers, iPod dock extenders, earbud cases and Mini DisplayPort adapters for Mac & PC.

## External links

- SendStation Systems Website [2]

## References

[1] http://www.sendstation.com/
[2] http://www.sendstation.com

# iPod

The iPod line as of September 2010; from left to right: iPod shuffle, iPod nano, iPod classic, iPod touch.

| | |
|---|---|
| **Developer** | Apple Inc. |
| **Manufacturer** | Foxconn (OEM)[1] |
| **Type** | Portable Media Player (PMP) |
| **Retail availability** | October 23, 2001 – present (first launched) |
| **Units sold** | Over 300,000,000 worldwide as of October 2011 (see chart below) |
| **CPU** | Samsung ARM and Apple A4 |
| **Online services** | iTunes Music Store, App Store, iCloud, iBooks, MobileMe, Game Center<br>(online services available only on iPod touch) |

**iPod** is a line of portable media players created and marketed by Apple Inc.. The product line-up consists of the hard drive-based iPod classic, the touchscreen iPod touch, the compact iPod nano and the ultra-compact iPod shuffle. iPod classic models store media on an internal hard drive, while all other models use flash memory to enable their smaller size (the discontinued mini used a Microdrive miniature hard drive). As with many other digital music players, iPods can serve as external data storage devices. Storage capacity varies by model, ranging from 2 GB for the iPod shuffle to 160 GB for the iPod classic. The iPod line was announced by Apple on October 23, 2001, and released on November 10, 2001.

All of the models have been redesigned multiple times since their introduction. The most recent iPod redesigns were introduced on September 1, 2010. Apple's iTunes software can be used to transfer music to the devices from computers using certain versions of Apple Macintosh and Microsoft Windows operating systems.[2] For users who choose not to use iTunes or whose computers cannot run iTunes, several open source alternatives are available for the iPod.[3] iTunes and its alternatives may also transfer photos, videos, games, contact information, e-mail settings, Web bookmarks, and calendars to iPod models supporting those features. Discontinued models of the line include the iPod mini and the iPod photo, the former being replaced by the iPod nano, and the latter reintegrated into the main iPod line (now the iPod classic).

In iOS versions prior to iOS 5, the iPod branding is also used for the media player applications included with the iPhone and iPad; the iPhone version is essentially a combination of the Music and Videos apps on the iPod touch. As of iOS 5, separate apps named 'Music' and 'Video' are standardized across all iOS-powered products.[4] While the iPhone and iPad have essentially the same media-player capabilities as the iPod line, they are generally treated as separate products. In the last few years, iPhone and iPad sales have overtaken those of the iPod.

## History and design

The iPod line came from Apple's "digital hub" category,[5] when the company began creating software for the growing market of personal digital devices. Digital cameras, camcorders and organizers had well-established mainstream markets, but the company found existing digital music players "big and clunky or small and useless" with user interfaces that were "unbelievably awful,"[5] so Apple decided to develop its own. As ordered by CEO Steve Jobs, Apple's hardware engineering chief Jon Rubinstein assembled a team of engineers to design the iPod line, including hardware engineers Tony Fadell and Michael Dhuey,[6] and design engineer Jonathan Ive.[5] The product was developed in less than one year and unveiled on October 23, 2001. Jobs announced it as a Mac-compatible product with a 5 GB hard drive that put "1,000 songs in your pocket."[7]

Various iPod models, all of which have been discontinued or updated.

Apple did not develop the iPod software entirely in-house, instead using PortalPlayer's reference platform based on two ARM cores. The platform had rudimentary software running on a commercial microkernel embedded operating system. PortalPlayer had previously been working on an IBM-branded MP3 player with Bluetooth headphones.[5] Apple contracted another company, Pixo, to help design and implement the user interface under the direct supervision of Steve Jobs.[5] As development progressed, Apple continued to refine the software's look and feel. Starting with the iPod mini, the Chicago font was replaced with Espy Sans. Later iPods switched fonts again to Podium Sans—a font similar to Apple's corporate font, Myriad. iPods with color displays then adopted some Mac OS X themes like Aqua progress bars, and brushed metal meant to evoke a combination lock. In 2007, Apple modified the iPod interface again with the introduction of the sixth-generation iPod classic and third-generation iPod nano by changing the font to Helvetica and, in most cases, splitting the screen in half by displaying the menus on the left and album artwork, photos, or videos on the right (whichever was appropriate for the selected item).

In September 2007, during a lawsuit with patent holding company Burst.com, Apple drew attention to a patent for a similar device that was developed in 1979. Kane Kramer applied for a UK patent for his design of a "plastic music box" in 1981, which he called the IXI.[8] He was unable to secure funding to renew the US$ 120,000 worldwide patent, so it lapsed and Kramer never profited from his idea.[8]

### Trademark

The name *iPod* was proposed by Vinnie Chieco, a freelance copywriter, who (with others) was called by Apple to figure out how to introduce the new player to the public. After Chieco saw a prototype, he thought of the movie *2001: A Space Odyssey* and the phrase "Open the pod bay door, Hal!", which refers to the white EVA Pods of the Discovery One spaceship. Chieco saw an analogy to the relationship between the spaceship and the smaller independent pods in the relationship between a personal computer and the music player.[5] Apple researched the trademark and found that it was already in use. Joseph N. Grasso of New Jersey had originally listed an "iPod" trademark with the U.S. Patent and Trademark Office (USPTO) in July 2000 for Internet kiosks. The first iPod kiosks had been demonstrated to the public in New Jersey in March 1998, and commercial use began in January 2000, but had apparently been discontinued by 2001. The trademark was registered by the USPTO in November 2003, and Grasso assigned it to Apple Computer, Inc. in 2005.[9]

The earliest recorded use in commerce of an "iPod" trademark was in 1991 by Chrysalis Corp. of Sturgis, Michigan, styled "iPOD".[10]

## Software

The iPod line can play several audio file formats including MP3, AAC/M4A, Protected AAC, AIFF, WAV, Audible audiobook, and Apple Lossless. The iPod photo introduced the ability to display JPEG, BMP, GIF, TIFF, and PNG image file formats. Fifth and sixth generation iPod classics, as well as third generation iPod nanos, can additionally play MPEG-4 (H.264/MPEG-4 AVC) and QuickTime video formats, with restrictions on video dimensions, encoding techniques and data-rates.[11] Originally, iPod software only worked with Mac OS; iPod software for Microsoft Windows was launched with the second generation model.[12] Unlike most other media players, Apple does not support Microsoft's WMA audio format—but a converter for WMA files without Digital Rights Management (DRM) is provided with the Windows version of iTunes. MIDI files also cannot be played, but can be converted to audio files using the "Advanced" menu in iTunes. Alternative open-source audio formats, such as Ogg Vorbis and FLAC, are not supported without installing custom firmware onto an iPod (e.g., Rockbox).

During installation, an iPod is associated with one host computer. Each time an iPod connects to its host computer, iTunes can synchronize entire music libraries or music playlists either automatically or manually. Song ratings can be set on an iPod and synchronized later to the iTunes library, and vice versa. A user can access, play, and add music on a second computer if an iPod is set to manual and not automatic sync, but anything added or edited will be reversed upon connecting and syncing with the main computer and its library. If a user wishes to automatically sync music with another computer, an iPod's library will be entirely wiped and replaced with the other computer's library.

### User interface

The signature iPod click wheel.

iPods with color displays use anti-aliased graphics and text, with sliding animations. All iPods (except the 3rd-generation iPod shuffle, the 6th generation iPod nano, and iPod touch) have five buttons and the later generations have the buttons integrated into the click wheel – an innovation that gives an uncluttered, minimalist interface. The buttons perform basic functions such as menu, play, pause, next track, and previous track. Other operations, such as scrolling through menu items and controlling the volume, are performed by using the click wheel in a rotational manner. The 3rd-generation iPod shuffle does not have any controls on the actual player; instead it has a small control on the earphone cable, with volume-up and -down buttons and a single button for play and pause, next track, etc. The iPod touch has no click-wheel; instead it uses a 3.5" touch screen along with a home button, sleep/wake button and (on the second and third generations of the iPod touch) volume-up and -down buttons. The user interface for the iPod touch is identical to that of the iPhone. Differences include a lack of a phone application. Both devices use iOS.

### iTunes Store

The iTunes Store (introduced April 29, 2003) is an online media store run by Apple and accessed through iTunes. The store became the market leader soon after its launch[13] and Apple announced the sale of videos through the store on October 12, 2005. Full-length movies became available on September 12, 2006.[14]

At the time the store was introduced, purchased audio files used the AAC format with added encryption, based on the FairPlay DRM system. Up to five authorized computers and an unlimited number of iPods could play the files. Burning the files with iTunes as an audio CD, then re-importing would create music files without the DRM. The DRM could also be removed using third-party software. However, in a deal with Apple, EMI began selling DRM-free, higher-quality songs on the iTunes Stores, in a category called "iTunes Plus." While individual songs were made available at a cost of US$1.29, 30¢ more than the cost of a regular DRM song, entire albums were available for the same price, US$9.99, as DRM encoded albums. On 17 October 2007, Apple lowered the cost of individual iTunes Plus songs to US$0.99 per song, the same as DRM encoded tracks. On January 6, 2009, Apple announced that DRM has been removed from 80% of the music catalog, and that it would be removed from all music by April 2009.

iPods cannot play music files from competing music stores that use rival-DRM technologies like Microsoft's protected WMA or RealNetworks' Helix DRM. Example stores include Napster and MSN Music. RealNetworks claims that Apple is creating problems for itself[15] by using FairPlay to lock users into using the iTunes Store. Steve Jobs stated that Apple makes little profit from song sales, although Apple uses the store to promote iPod sales.[16] However, iPods can also play music files from online stores that do not use DRM, such as eMusic or Amie Street.

Universal Music Group decided not to renew their contract with the iTunes Music Store on July 3, 2007. Universal will now supply iTunes in an 'at will' capacity.[17]

Apple debuted the iTunes Wi-Fi Music Store on September 5, 2007, in its Media Event entitled "The Beat Goes On..." This service allows users to access the Music Store from either an iPhone or an iPod touch and download songs directly to the device that can be synced to the user's iTunes Library over a WiFi connection, or, in the case of a iPhone, the telephone network.

### Games

Video games are playable on various versions of iPods. The original iPod had the game *Brick* (originally invented by Apple's co-founder Steve Wozniak) included as an easter egg hidden feature; later firmware versions added it as a menu option. Later revisions of the iPod added three more games: *Parachute*, *Solitaire*, and *Music Quiz*.

In September 2006, the iTunes Store began to offer additional games for purchase with the launch of iTunes 7, compatible with the fifth generation iPod with iPod software 1.2 or later. Those games were: *Bejeweled*, *Cubis 2*, *Mahjong*, *Mini Golf*, *Pac-Man*, *Tetris*, *Texas Hold 'Em*, *Vortex*, *Asphalt 4: Elite Racing* and *Zuma*. Additional games have since been added. These games work on the 6th and 5th generation iPod classic and the 5th and 4th generation iPod nano.

With third parties like Namco, Square Enix, Electronic Arts, Sega, and Hudson Soft all making games for the iPod, Apple's MP3 player has taken steps towards entering the video game handheld console market. Even video game magazines like GamePro and EGM have reviewed and rated most of their games as of late.[18]

The games are in the form of .ipg files, which are actually .zip archives in disguise. When unzipped, they reveal executable files along with common audio and image files, leading to the possibility of third party games. Apple has not publicly released a software development kit (SDK) for iPod-specific development.[19] Apps produced with the iPhone SDK are compatible only with the iOS on the iPod touch and iPhone, which cannot run clickwheel-based games.

## File storage and transfer

All iPods except for the iPod touch can function in "disk mode" as mass storage devices to store data files[20] but this may not be the default behavior, and in the case of the iPod touch, requires special software. If an iPod is formatted on a Mac OS X computer, it uses the HFS+ file system format, which allows it to serve as a boot disk for a Mac computer.[21] If it is formatted on Windows, the FAT32 format is used. With the release of the Windows-compatible iPod, the default file system used on the iPod line switched from HFS+ to FAT32, although it can be reformatted to either file system (excluding the iPod shuffle which is strictly FAT32). Generally, if a new iPod (excluding the iPod shuffle) is initially plugged into a computer running Windows, it will be formatted with FAT32, and if initially plugged into a Mac running Mac OS X it will be formatted with HFS+.[22]

Unlike many other MP3 players, simply copying audio or video files to the drive with a typical file management application will not allow an iPod to properly access them. The user must use software that has been specifically designed to transfer media files to iPods, so that the files are playable and viewable. Usually *iTunes* is used to transfer media to an iPod, though several alternative third-party applications are available on a number of different platforms.

*iTunes* 7 and above can transfer purchased media of the iTunes Store from an iPod to a computer, provided that computer containing the DRM protected media is authorized to play it.

Media files are stored on an iPod in a hidden folder, along with a proprietary database file. The hidden content can be accessed on the host operating system by enabling hidden files to be shown. The media files can then be recovered manually by copying the files or folders off the iPod. Many third-party applications also allow easy copying of media files off of an iPod.

# Hardware

## Chipsets and Electronics

| Chipset or Electronic | Product(s) | Component(s) |
|---|---|---|
| Microcontroller | iPod classic first to third generations | Two ARM 7TDMI-derived CPUs running at 90 MHz |
| | iPod fourth and fifth generations, iPod mini, iPod nano first generation | Variable-speed ARM 7TDMI CPUs, running at a peak of 80 MHz to save battery life |
| | iPod nano second generation | Samsung System-on-a-chip, based around an ARM processor.[23] [24] |
| | iPod shuffle first generation | SigmaTel STMP3550 chip that handles both the music decoding and the audio circuitry.[25] |
| Audio Chip | All iPods (except the iPod shuffle, 6G classic and 2G touch)[26] | Audio Codecs developed by Wolfson Microelectronics |
| | Sixth-generation iPod classic | Cirrus Logic Audio Codec Chip |
| Storage Medium | iPod classic | 45.7 mm (1.8 in) hard drives (ATA-6, 4200 rpm with proprietary connectors) made by Toshiba |
| | iPod mini | 25.4 mm (1 in) Microdrive by Hitachi and Seagate |
| | iPod nano | Flash Memory from Samsung, Toshiba, and others |
| | iPod shuffle and touch | Flash Memory |
| Batteries | iPod first and second generation, shuffle | Internal Lithium Polymer Batteries |
| | iPod third generation onward, iPod mini, iPod nano, iPod touch, fourth generation iPod shuffle (including maybe earlier) | Internal Lithium-Ion Batteries |
| Display | iPod nano | 1.54-inch (diagonal) Multi-Touch, 240-by-240 resolution at 220 pixels per inch[27] |
| | iPod classic | 2.5-inch (diagonal) color LCD with LED backlight, 320-by-240 resolution at 163 pixels per inch[28] |
| | iPod touch | 3.5-inch (diagonal) widescreen Multi-Touch, 960-by-640 resolution at 326 pixels per inch[29] |

## Connectivity

See also: iPod dock connector

Four iPod wall chargers for North America, all made by Apple. These have FireWire (left) and USB (right three) connectors, which allow iPods to charge without a computer. The units have been miniaturized over time.

Originally, a FireWire connection to the host computer was used to update songs or recharge the battery. The battery could also be charged with a power adapter that was included with the first four generations. The third generation began including a 30-pin dock connector, allowing for FireWire or USB connectivity. This provided better compatibility with non-Apple machines, as most of them did not have FireWire ports at the time. Eventually Apple began shipping iPods with USB cables instead of FireWire, although the latter was available separately. As of the first-generation iPod nano and the fifth-generation iPod classic, Apple discontinued using FireWire for data transfer (while still allowing for use of FireWire to charge the device) in an attempt to reduce cost and form factor. As of the second-generation iPod touch and the fourth-generation iPod nano, FireWire charging ability has been removed. The second-, third-, and fourth-generation iPod shuffle uses a single 3.5 mm jack which acts as both a headphone jack and a data port for the dock.

The dock connector also allowed the iPod to connect to accessories, which often supplement the iPod's music, video, and photo playback. Apple sells a few accessories, such as the now-discontinued iPod Hi-Fi, but most are manufactured by third parties such as Belkin and Griffin. Some peripherals use their own interface, while others use the iPod's own screen. Because the dock connector is a proprietary interface, the implementation of the interface requires paying royalties to Apple.[30]

## Accessories

Made for iPod

The "Made for iPod" logo found on most classic iPod accessories

Many accessories have been made for the iPod line. A large number are made by third party companies, although many, such as the iPod Hi-Fi, are made by Apple. Some accessories add extra features that other music players have, such as sound recorders, FM radio tuners, wired remote controls, and audio/visual cables for TV connections. Other accessories offer unique features like the Nike+iPod pedometer and the iPod Camera Connector. Other notable accessories include external speakers, wireless remote controls, protective case, screen films, and wireless earphones.[31] Among the first accessory manufacturers were Griffin Technology, Belkin, JBL, Bose, Monster Cable, and SendStation.

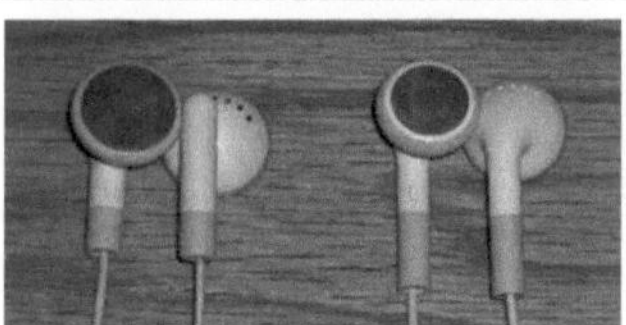

Two designs of iPod earphones. The current version is shown on the right.

BMW released the first iPod automobile interface,[32] allowing drivers of newer BMW vehicles to control an iPod using either the built-in steering wheel controls or the radio head-unit buttons. Apple announced in 2005 that similar systems would be available for other vehicle brands, including Mercedes-Benz,[33] Volvo,[34] Nissan, Toyota,[35] Alfa Romeo, Ferrari,[36] Acura, Audi, Honda,[37] Renault, Infiniti[38] and Volkswagen.[39] Scion offers standard iPod connectivity on all their cars.

Some independent stereo manufacturers including JVC, Pioneer, Kenwood, Alpine, Sony, and Harman Kardon also have iPod-specific integration solutions. Alternative connection methods include adapter kits (that use the cassette deck or the CD changer port), audio input jacks, and FM transmitters such as the iTrip—although personal FM transmitters are illegal in some countries. Many car manufacturers have added audio input jacks as standard.[40]

Beginning in mid-2007, four major airlines, United, Continental, Delta, and Emirates, reached agreements to install iPod seat connections. The free service will allow passengers to power and charge an iPod, and view video and music libraries on individual seat-back displays.[41] Originally KLM and Air France were reported to be part of the deal with Apple, but they later released statements explaining that they were only contemplating the possibility of incorporating such systems.[42]

### Audio performance

The third-generation iPod had a weak bass response, as shown in audio tests.[43] [44] The combination of the undersized DC-blocking capacitors and the typical low-impedance of most consumer headphones form a high-pass filter, which attenuates the low-frequency bass output. Similar capacitors were used in the fourth-generation iPods.[45] The problem is reduced when using high-impedance headphones and is completely masked when driving high-impedance (line level) loads, such as an external headphone amplifier. The first-generation iPod shuffle uses a dual-transistor output stage,[43] rather than a single capacitor-coupled output, and does not exhibit reduced bass response for any load.

For all iPods released in 2006 and earlier, some equalizer (EQ) sound settings could "distort the [bass] sound far too easily, even on undemanding songs".[46] This would happen for EQ settings like R&B, Rock, Acoustic, and Bass Booster, because the equalizer amplified the digital audio level beyond the software's limit, causing distortion (clipping) on bass instruments.

From the fifth-generation iPod on, Apple introduced a user-configurable volume limit in response to concerns about hearing loss.[47] Users report that in the sixth-generation iPod, the maximum volume output level is limited to 100 dB in EU markets. Apple previously had to remove iPods from shelves in France for exceeding this legal limit.[48]

## Models

While the suffix "classic" was not introduced until the sixth generation, it has been applied here retroactively to all generic iPods for clarity.

| Model | Generation | Image | Capacity | Connection | Original release date | Minimum OS to sync | Rated battery life (hours) | | |
|---|---|---|---|---|---|---|---|---|---|

| | | | | | | | | | |
|---|---|---|---|---|---|---|---|---|---|
| **Classic** | **first** |  | 5, 10 GB | FireWire | 23 October 2001 | Mac: 9, 10.1 | *audio*: 10 | | |
| | | | First model, with mechanical scroll wheel. 10 GB model released later. | | | | | | |
| | **second** |  | 10, 20 GB | FireWire | 17 July 2002 | Mac: 10.1<br>Win: 2000 | *audio*: 10 | | |
| | | | Touch-sensitive wheel. FireWire port had a cover. Hold switch revised. Windows compatibility through Musicmatch. | | | | | | |
| | **third** |  | 10, 15, 20, 30, 40 GB | FireWire (USB for syncing only) | 28 April 2003 | Mac: 10.1<br>Win: 2000 | *audio*: 8 | | |
| | | | First complete redesign with all-touch interface, dock connector, and slimmer case. Musicmatch support dropped with later release of iTunes 4.1 for Windows. | | | | | | |
| | **fourth (Photo) (Color)** |  | 20, 40 GB | FireWire or USB | 19 July 2004 | Mac: 10.2<br>Win: 2000 | *audio*: 12 | | |
| | | | Adopted Click Wheel from iPod Mini, hold switch redesigned. | | | | | | |
| | |  | **photo**:<br>30, 40, 60 GB | FireWire or USB | 26 October 2004 | Mac: 10.2<br>Win: 2000 | *audio*: 15<br>*slideshow*: 5 | | |
| | | | **color**:<br>20, 60 GB | | 28 June 2005 | | | | |
| | | | Premium spin-off of 4G iPod with color screen and picture viewing. *Later re-integrated into main iPod line.* | | | | | | |
| | **fifth** |  | 30, 60, 80 GB | USB (FireWire for charging only) | 12 October 2005 | Mac: 10.3<br>Win: 2000 | **30 GB**<br>*audio*: 14<br>*video*: 2<br>(later 3.5) | **60/80 GB**<br>*audio*: 20<br>*video*: 3/6.5 | |
| | | | Second full redesign with a slimmer case, and larger screen with video playback. Offered in black or white. Hardware and firmware updated with 60 GB model replaced with 80 GB model on 12 September 2006. | | | | | | |
| | **sixth** |  | 80, 120, 160 GB | USB (FireWire for charging only) | 5 September 2007 | Mac: 10.4<br>Win: XP | **80 GB**<br>*audio*: 30<br>*video*: 5 | **120 GB**<br>*audio*: 36<br>*video*: 6 | **160 GB**<br>2007 model<br>*audio*: 40<br>*video*: 7<br>2009 model<br>*audio*: 36<br>*video*: 6 |
| | | | Introduced the "Classic" suffix. New interface and anodized aluminum front plate. Silver replaces white. In September 2008 the hardware and firmware was updated with a 120 GB model replacing the 80 GB model and the 160 GB model was discontinued. In September 2009, the 120GB model was replaced with a 160GB model. | | | | | | |
| **Mini** | **first** |  | 4 GB | USB or FireWire | 6 January 2004 | Mac: 10.1<br>Win: 2000 | *audio*: 8 | | |
| | | | New smaller model, available in 5 colors. Introduced the "Click Wheel". | | | | | | |
| | **second** |  | 4, 6 GB | USB or FireWire | 22 February 2005 | Mac: 10.2<br>Win: 2000 | *audio*: 18 | | |
| | | | Brighter color variants with longer battery life. Click Wheel lettering matched body color. Gold color discontinued. *Later replaced by iPod Nano.* | | | | | | |

| | | | | | | | | | |
|---|---|---|---|---|---|---|---|---|---|
| **Nano** | **first** |  | 1, 2, 4 GB | USB (FireWire for charging only) | 7 September 2005 | Mac: 10.3 Win: 2000 | *audio*: 14 *slideshow*: 4 | | |
| | | | *Replaced Mini.* Available in black or white and used flash memory. Color screen for picture viewing. 1 GB version released later. | | | | | | |
| | **second** |  | 2, 4, 8 GB | USB (FireWire for charging only) | 12 September 2006 | Mac: 10.3 Win: 2000 | *audio*: 24 *slideshow*: 5 | | |
| | | | Anodized aluminum casing and 6 colors available. | | | | | | |
| | **third** |  | 4, 8 GB | USB (FireWire for charging only) | 5 September 2007 | Mac: 10.4 Win: XP | *audio*: 24 *video*: 5 | | |
| | | | 2" QVGA screen, colors refreshed with chrome back, new interface, video capability, smaller Click Wheel. | | | | | | |
| | **fourth** |  | 4, 8, 16 GB | USB | 9 September 2008 | Mac: 10.4 Win: XP | *audio*: 24 *video*: 4 | | |
| | | | Revert to tall form and all-aluminum enclosure with 9 color choices, added accelerometer for shake and horizontal viewing. 4 GB model limited release in select markets. | | | | | | |
| | **fifth** |  | 8, 16 GB | USB | 9 September 2009 | Mac: 10.4 Win: XP | *audio*: 24 *video*: 5 | | |
| | | | First iPod to include a video camera; also included a larger screen, an FM radio, a speaker, a pedometer, and a polished exterior case while retaining the similar colors as the fourth generation model. | | | | | | |
| | **sixth** |  | 8, 16 GB | USB | 1 September 2010 | Mac: 10.5 Win: XP | *audio*: 24 | | |
| | | | First iPod nano to include multi-touch screen; clip from iPod shuffle added. Video playback, speakers and camera removed. | | | | | | |
| **Shuffle** | **first** |  | 512 MB, 1 GB | USB(no adaptor required) | 11 January 2005 | Mac: 10.2 Win: 2000 | *audio*: 12 | | |
| | | | New entry-level model. Uses flash memory and has no screen. | | | | | | |
| | **second** |  | 1, 2 GB | USB | 12 September 2006 | Mac: 10.3 Win: 2000 | *audio*: 12 | | |
| | | | Smaller clip design with anodized aluminum casing. 4 color options added later. Colors were later refreshed twice. | | | | | | |
| | **third** |  | 2, 4 GB | USB | 11 March 2009 | Mac: 10.4 Win: XP | *audio*: 10 | | |
| | | | Smaller design with controls relocated to right earbud cable. Introduced with two colors, and features VoiceOver. More colors and 2GB model added in September 2009. | | | | | | |
| | **fourth** |  | 2 GB | USB | 1 September 2010 | Mac: 10.5 Win: XP | *audio*: 15 | | |
| | | | Controls returned to the body of the iPod. Introduced with five colors, and features VoiceOver. | | | | | | |

| | | | | | | | | | |
|---|---|---|---|---|---|---|---|---|---|
| **Touch** | **first** |  | 8, 16, 32 GB | USB (FireWire for charging only)[49] | 5 September 2007 | Mac: 10.4 Win: XP | *audio*: 22 *video*: 5 | | |
| | | | First iPod with Wi-Fi and a Multi-Touch interface. Features Safari browser and wireless access to the iTunes Store and YouTube. 32 GB model later added. iOS 2.0 and App Store access requires an upgrade fee. | | | | | | |
| | **second** |  | 8, 16, 32 GB | USB | 9 September 2008 | Mac: 10.4 Win: XP | *audio*: 36 *video*: 6 | | |
| | | | New tapered chrome back with Nike+ functionality, volume buttons, and built-in speaker added. iOS 2.0 and App Store access standard. Bluetooth support added but not made active until iOS 3.0, which requires an upgrade fee. | | | | | | |
| | **third** | | 32, 64 GB | USB | 9 September 2009 | Mac: 10.4 Win: XP | *audio*: 30 *video*: 6 | | |
| | | | Updated to include the upgraded internals from the iPhone 3GS; includes Voice Control support and bundled remote earphones. | | | | | | |
| | **fourth** |  | 8, 32, 64 GB | USB | 9 September 2010 | Mac: 10.5 Win: XP | *audio*: 40 *video*: 7 | | |
| | | | New thinner design including two cameras for FaceTime and HD video recording, hold button moved to top right corner, Retina display similar to iPhone 4, Apple A4 chip. White-colored version added on 4 October 2011. | | | | | | |
| *Sources: Apple Inc.*[50]*, Mactracker*[51] | | | | | | | | | |

## Timeline of iPod models

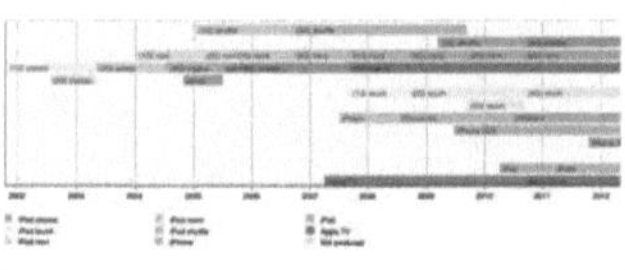

*Sources: Apple press release library,*[52] *Mactracker Apple Inc. model database*[51]

## Patent disputes

In 2005, Apple faced two lawsuits claiming patent infringement by the iPod line and its associated technologies:[53] Advanced Audio Devices claimed the iPod line breached its patent on a "music jukebox",[54] while a Hong Kong-based IP portfolio company called Pat-rights filed a suit claiming that Apple's FairPlay technology breached a patent[55] issued to inventor Ho Keung Tse. The latter case also includes the online music stores of Sony, RealNetworks, Napster, and Musicmatch as defendants.[56]

Apple's application to the United States Patent and Trademark Office for a patent on "rotational user inputs",[57] as used on the iPod interface, received a third "non-final rejection" (NFR) in August 2005. Also in August 2005, Creative Technology, one of Apple's main rivals in the MP3 player market, announced that it held a patent[58] on part of the music selection interface used by the iPod line, which Creative dubbed the "Zen Patent", granted on August 9, 2005.[59] On May 15, 2006, Creative filed another suit against Apple with the United States District Court for the Northern District of California. Creative also asked the United States International Trade Commission to investigate whether Apple was breaching U.S. trade laws by importing iPods into the United States.[60]

On August 24, 2006, Apple and Creative announced a broad settlement to end their legal disputes. Apple will pay Creative US$100 million for a paid-up license, to use Creative's awarded patent in all Apple products. As part of the agreement, Apple will recoup part of its payment, if Creative is successful in licensing the patent. Creative then announced its intention to produce iPod accessories by joining the *Made for iPod* program.[61]

## Sales

Since October 2004, the iPod line has dominated digital music player sales in the United States, with over 90% of the market for hard drive-based players and over 70% of the market for all types of players.[62] During the year from January 2004 to January 2005, the high rate of sales caused its U.S. market share to increase from 31% to 65% and in July 2005, this market share was measured at 74%. In January 2007 the iPod market share reached 72.7% according to Bloomberg Online.

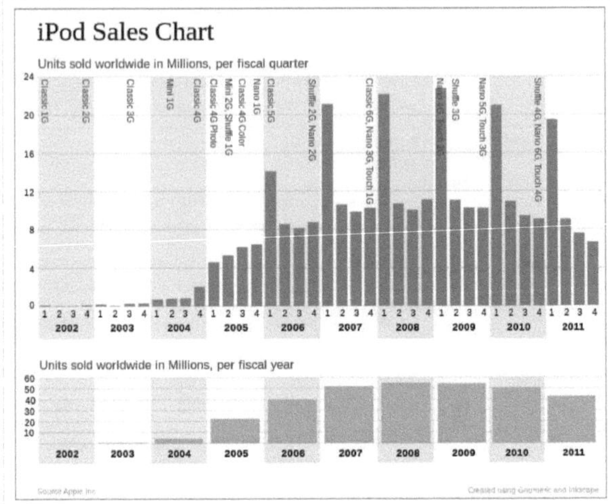


iPod quarterly sales. Click for table of data and sources. Note that Q1 is October through December of previous year, the holiday season.

On January 8, 2004, Hewlett-Packard (HP) announced that they would sell HP-branded iPods under a license agreement from Apple. Several new retail channels were used—including Wal-Mart—and these iPods eventually made up 5% of all iPod sales. In July 2005, HP stopped selling iPods due to unfavorable terms and conditions imposed by Apple.[63]

In January 2007, Apple reported record quarterly revenue of US$7.1 billion, of which 48% was made from iPod sales.[64]

On April 9, 2007, it was announced that Apple had sold its one-hundred millionth iPod, making it the biggest selling digital music player of all time. In April 2007, Apple reported second quarter revenue of US$5.2 billion, of which 32% was made from iPod sales.[65] Apple and several industry analysts suggest that iPod users are likely to purchase other Apple products such as Mac computers.[66]

On October 22, 2007, Apple reported quarterly revenue of US$6.22 billion, of which 30.69% came from Apple notebook sales, 19.22% from desktop sales and 26% from iPod sales. Apple's 2007 year revenue increased to US$24.01 billion with US$3.5 billion in profits. Apple ended the fiscal year 2007 with US$15.4 billion in cash and no debt.[67]

On January 22, 2008, Apple reported the best quarter revenue and earnings in Apple's history so far. Apple posted record revenue of US$9.6 billion and record net quarterly profit of US$1.58 billion. 42% of Apple's revenue for the First fiscal quarter of 2008 came from iPod sales, followed by 21% from notebook sales and 16% from desktop sales.[68]

On October 21, 2008, Apple reported that only 14.21% of total revenue for fiscal quarter 4 of year 2008 came from iPods.[69] At the September 9, 2009 keynote presentation at the Apple Event, Phil Schiller announced total cumulative sales of iPods exceeded 220 million.[70]

As of October 2011, Apple reported that total number of iPods sold worldwide was 300 million.[71]

## Industry impact

iPods have won several awards ranging from engineering excellence,[72] to most innovative audio product,[73] to fourth best computer product of 2006.[74] iPods often receive favorable reviews; scoring on looks, clean design, and ease of use. PC World says that iPod line has "altered the landscape for portable audio players".[73] Several industries are modifying their products to work better with both the iPod line and the AAC audio format. Examples include CD copy-protection schemes,[75] and mobile phones, such as phones from Sony Ericsson and Nokia, which play AAC files rather than WMA.

Besides earning a reputation as a respected entertainment device, the iPod has also been accepted as a business device. Government departments, major institutions and international organisations have turned to the iPod line as a delivery mechanism for business communication and training, such as the Royal and Western Infirmaries in Glasgow, Scotland, where iPods are used to train new staff.[76]

iPods have also gained popularity for use in education. Apple offers more information on educational uses for iPods on their website,[77] including a collection of lesson plans. There has also been academic research done in this area in nursing education[78] and more general K-16 education.[79] Duke University provided iPods to all incoming freshmen in the fall of 2004, and the iPod program continues today with modifications.[80] *Entertainment Weekly* put it on its end-of-the-decade, "best-of" list, saying, "Yes, children, there really was a time when we roamed the earth without thousands of our favorite jams tucked comfortably into our hip pockets. Weird."[81]

## Criticism

### Battery problems

The advertised battery life on most models is different from the real-world achievable life. For example, the fifth generation 30 GB iPod is advertised as having up to 14 hours of music playback. An MP3.com report stated that this was virtually unachievable under real-life usage conditions, with a writer for MP3.com getting on average less than 8 hours from an iPod.[82] In 2003, class action lawsuits were brought against Apple complaining that the battery charges lasted for shorter lengths of time than stated and that the battery degraded over time.[83] The lawsuits were settled by offering individuals either US$50 store credit or a free battery replacement.[84]

iPod batteries are not designed to be removed or replaced by the user, although some users have been able to open the case themselves, usually following instructions from third-party vendors of iPod replacement batteries. Compounding the problem, Apple initially would not replace worn-out batteries. The official policy was that the customer should buy a refurbished replacement iPod, at a cost almost equivalent to a brand new one. All lithium-ion batteries lose capacity during their lifetime even when not in use[85] (guidelines are available for prolonging life-span) and this situation led to a market for third-party battery replacement kits.

Apple announced a battery replacement program on November 14, 2003, a week before[86] a high publicity stunt and website by the Neistat Brothers.[87] The initial cost was US$99,[88] and it was lowered to US$59 in 2005. One week later, Apple offered an extended iPod warranty for US$59.[89] For the iPod nano, soldering tools are needed because the battery is soldered onto the main board. Fifth generation iPods have their battery attached to the backplate with adhesive.[90] [91]

The first generation iPod nano may overheat and pose a health and safety risk. Affected iPod nanos were sold between September 2005 and December 2006. This is due to a flawed battery used by Apple from a single battery manufacturer.[92] Apple recommended that owners of affected iPod Nanos stop using them. Under an Apple product replacement program, affected Nanos were replaced with current generation Nanos free of charge.

### Reliability and durability

iPods have been criticized for alleged short life-span and fragile hard drives. A 2005 survey conducted on the MacInTouch website found that the iPod line had an average failure rate of 13.7% (although they note that comments from respondents indicate that "the true iPod failure rate may be lower than it appears"). It concluded that some models were more durable than others.[93] In particular, failure rates for iPods employing hard drives was usually above 20% while those with flash memory had a failure rate below 10%. In late 2005, many users complained that the surface of the first generation iPod nano can become scratched easily, rendering the screen unusable.[94] [95] A class action lawsuit was also filed.[96] Apple initially considered the issue a minor defect, but later began shipping these iPods with protective sleeves.

### Allegations of worker exploitation

On June 11, 2006, the British tabloid *The Mail on Sunday* reported that iPods are mainly manufactured by workers who earn no more than US$50 per month and work 15-hour shifts.[97] Apple investigated the case with independent auditors and found that, while some of the plant's labour practices met Apple's Code of Conduct, others did not: employees worked over 60 hours a week for 35% of the time, and worked more than six consecutive days for 25% of the time.[98]

Foxconn, Apple's manufacturer, initially denied the abuses,[99] but when an auditing team from Apple found that workers had been working longer hours than were allowed under Chinese law, they promised to prevent workers working more hours than the code allowed. Apple hired a workplace standards auditing company, Verité, and joined the Electronic Industry Code of Conduct Implementation Group to oversee the measures. On December 31, 2006, workers at the Foxconn factory in Longhua, Shenzhen formed a union affiliated with the All-China Federation of Trade Unions,[100] the Chinese government-approved union umbrella organization.[101] [102]

In 2010, a number of workers committed suicide at a Foxconn operations in China. Apple, HP, and others stated that they were investigating the situation. Foxconn guards have been videotaped beating employees. Another employee killed himself in 2009 when an Apple prototype went missing, and claimed in messages to friends, that he had been beaten and interrogated.[103] [104]

As of 2006, the iPod was produced by about 14,000 workers in the U.S. and 27,000 overseas. Further, the salaries attributed to this product were overwhelmingly distributed to highly skilled U.S. professionals, as opposed to lower skilled U.S. retail employees or overseas manufacturing labor. One interpretation of this result is that U.S. innovation can create more jobs overseas than domestically.[105]

## See also

- Comparison of portable media players
- Comparison of iPod managers
- iPhone

## References

[1] "Apple, HP, Dell Looking Into Foxconn's Working Conditions In China – ChinaTechNews.com – The Technology Source for the Latest Chinese News on Internet, Computers, Digital, Science, Electronics, Law, Security, Software, Web 2.0, Telecom, and Wireless Industries" (http://www.chinatechnews.com/2010/05/28/12117-apple-hp-dell-looking-into-foxconns-working-conditions-in-china). ChinaTechNews.com. May 28, 2010. . Retrieved June 2, 2010.

[2] Apple Inc.. "iTunes system requirements. Apple iTunes software runs on Mac OS X 10.3.9 or OS X 10.4.9 or later and on Microsoft Windows XP (Service Pack 2) or Vista" (http://www.apple.com/itunes/download/). . Retrieved May 28, 2008.

[3] McKillop, Ross (July 8, 2007). "10 Alternatives to iTunes for managing your iPod" (http://www.simplehelp.net/2007/07/08/10-alternatives-to-itunes-for-managing-your-ipod/). Simple Help. . Retrieved May 28, 2008.

[4] "APPLE IOS 5 IPHONE / IPAD HANDS-ON (screenshot of iPhone home screen with iOS 5 preview)" (http://www.bgr.com/nggallery/page-320/album-all/gallery-186/image/2403/). *Boy Genius Report*. June 6, 2011. . Retrieved June 6, 2011.

[5] Kahney, Leander. Straight Dope on the iPod's Birth (http://www.wired.com/news/columns/cultofmac/0,71956-0.html), *Wired News*, October 17, 2006. Retrieved on October 30, 2006.

[6] " 2007 Engineer of the Year Finalist Michael Dhuey's Hardware Knowledge Helps Breathe Life Into iPod, TelePresence (http://www.designnews.com/article/CA6478178.html)", *Design News*, September 24, 2007.

[7] "A History of the iPod: 2000 to 2004" (http://lowendmac.com/orchard/05/origin-of-the-ipod.html). Lowendmac.com. . Retrieved 2011-08-15.

[8] Boffey, Daniel Apple admit Briton DID invent iPod, but he's still not getting any money (http://www.dailymail.co.uk/news/article-1053152/Apple-admit-Briton-DID-invent-iPod-hes-getting-money.html) *Daily Mail*, September 8, 2008. Retrieved on September 8, 2008.

[9] Serial No. 78018061, Registration No. 2781793, records of the U.S. Patent and Trademark Office. InPub, LLC, filed an "IPOD" trademark on June 1, 1999, for "computer software and hardware." The trademark was abandoned May 18, 2000, without commercial use.

[10] "Ipod by Chrysalis Incorporated" (http://www.trademarkia.com/ipod-74132201.html). Trademarkia. . Retrieved December 20, 2011.

[11] The restrictions vary from generation to generation; for the earliest video iPods, video is required to be Baseline Profile (BP), up to Level 1.3, meaning most significantly no B-frames (BP), a maximum bitrate of 768 kb/s (BP Level 1.3), and a maximum framerate of 30 frame/s at 320×240 resolution. Newer iPods support BP up to level 3.0 (10,000 kb/s), for a maximum framerate of 30 frame/s at 640×480 resolution. Current specifications can be seen at iPod classic Technical Specs (http://www.apple.com/ipodclassic/specs.html), and practical implementations can be seen in the libx264-ipod320.ffpreset and libx264-ipod640.ffpreset preset files for FFmpeg, as discussed in [Ffmpeg-user] Successful ipod h264 encoding (http://lists.mplayerhq.hu/pipermail/ffmpeg-user/2006-June/003218.html), by Daniel Rogers, Jun 11, 2006.

[12] "iTunes Downloads" (http://www.apple.com/itunes/download/). Apple.com. 2008-02-07. . Retrieved 2011-08-15.

[13] iTunes Music Store Catalog Tops One Million Songs (http://www.apple.com/pr/library/2004/aug/10itms.html), *Apple Inc.*, August 10, 2004. Retrieved on December 28, 2006.

[14] Scott-Joynt, Jeremy. Apple targets TV and film market (http://news.bbc.co.uk/1/hi/business/5339470.stm), *BBC News*, September 12, 2006. Retrieved on September 12, 2006.

[15] Kanellos, Michael (March 23, 2004). "Real's Glaser exhorts Apple to open iPod" (http://news.cnet.com/2100-1025_3-5177914.html). CNET News. . Retrieved December 20, 2011.

[16] Hansell, Saul (August 11, 2008). "The iTunes Store: Profit Machine" (http://bits.blogs.nytimes.com/2008/08/11/steve-jobs-tries-to-downplay-the-itunes-stores-profit/). August 11, 2008. . Retrieved December 20, 2011.

[17] Evans, Jonny. Universal confirms iTunes contract change (http://www.macworld.co.uk/ipod-itunes/news/index.cfm?newsid=18459), *Macworld UK*, July 4, 2007. Retrieved on July 5, 2007.

[18] "iPod Games Review Roundup" (http://web.archive.org/web/20110607151141/http://www.gamepro.com/article/features/81588/ipod-games-review-roundup/). Gamepro.com. 2006-09-15. Archived from the original (http://www.gamepro.com/article/features/81588/ipod-games-review-roundup/) on 2011-06-07. . Retrieved 2011-08-15.

[19] "What's Inside an iPod Game?" (http://www.bensinclair.com/article/whats-inside-an-ipod-game) bensinclair.com, September 14, 2006.

[20] "iPod touch: Appears in iTunes but not in Finder or Windows Desktop" (http://docs.info.apple.com/article.html?artnum=306414). Docs.info.apple.com. 2010-01-06. . Retrieved 2011-08-15.

[21] —Gina Trapani (2007-05-14). "How To: Boot up your Mac from your iPod" (http://lifehacker.com/software/how-to/boot-up-your-mac-from-your-ipod-260157.php). Lifehacker.com. . Retrieved 2011-08-15.

[22] "Restore iPod – Format iPod – Reformat iPod – iPod Restore Software" (http://ipod.about.com/od/restoreandreformatipod/a/ipod_format_a.htm). Ipod.about.com. February 1, 2010. . Retrieved June 2, 2010.

[23] Cassell, Jonathan. Apple Delivers More For Less With New iPod nano (http://www.isuppli.com/markech/default.asp?id=360), iSuppli Corporation, September 20, 2006. Retrieved on October 21, 2006.

[24] "mobile SoC" (http://web.archive.org/web/20080225041633/http://www.samsung.com/global/business/semiconductor/products/mobilesoc/Products_MobileSoC.html). Samsung Group. Archived from the original (http://www.samsung.com/global/business/semiconductor/products/mobilesoc/Products_MobileSoC.html) on February 25, 2008. . Retrieved August 4, 2009.

[25] Williams, Martyn. How Much Should an IPod shuffle Cost? (http://www.pcworld.com/article/119799-1/article.html), *PC World*, February 24, 2005. Retrieved on August 14, 2006.

[26] Macworld Wolfson loses Apple iPod business (http://www.macworld.co.uk/ipod-itunes/news/index.cfm?RSS&NewsID=20818)

[27] "iPod nano Technical Specifications" (http://www.apple.com/ipodnano/specs.html). Apple Inc.. . Retrieved September 3, 2010.

[28] "iPod classic Technical Specifications" (http://www.apple.com/ipodclassic/specs.html). Apple Inc.. . Retrieved August 4, 2009.

[29] "iPod touch Technical Specifications" (http://www.apple.com/ipodtouch/specs.html). Apple Inc.. . Retrieved September 3, 2010.

[30] Johnson, Joel (July 10, 2008). "How the "Apple Tax" Boosts Prices on iPod & iPhone Accessories" (http://www.popularmechanics.com/technology/industry/4272628.html?page=1). Popular Mechanics. . Retrieved September 9, 2008.

[31] In-The-Ear Bluetooth Earphones (http://www.etymotic.com/ephp/er88.aspx). Retrieved on February 17, 2007.

[32] iPod Your BMW (http://www.ipodyourbmw.com/). Retrieved on February 17, 2007.

[33] Apple & Mercedes-Benz Unveil iPod Integration Kit (http://www.apple.com/pr/library/2005/jan/11mercedes.html), *Apple Inc.*, January 11, 2005. Retrieved on June 20, 2006.

[34] Apple & Volvo Announce iPod Connectivity For Entire 2005 US Model Line (http://www.apple.com/pr/library/2005/jan/11volvo.html), *Apple Inc.*, January 11, 2005. Retrieved on June 20, 2006.

[35] "Toyota introduce iPod integration kit" (http://www.gizmag.com/go/7945/). Gizmag.com. September 4, 2007. . Retrieved June 2, 2010.

[36] Apple & Leading Car Companies Team Up to Deliver iPod Integration in 2005 (http://www.apple.com/pr/library/2005/jan/11cars.html), *Apple Inc.*, January 11, 2005. Retrieved on June 20, 2006.

[37] Honda Music Link for iPods (http://automobiles.honda.com/models/audio_accessories.asp?ModelName=Accord Coupe&PHOTO=0), *Honda*. Retrieved on February 17, 2007.

[38] Apple Car Integration page (http://www.apple.com/ipod/carintegration/#car_browser_infiniti)

[39] Apple Teams Up With Acura, Audi, Honda & Volkswagen to Deliver Seamless iPod Experience (http://www.apple.com/pr/library/2005/sep/07ipod_auto.html), *Apple Inc.*, September 7, 2005. Retrieved on June 20, 2006.

[40] Car Integration: iPod your car (http://www.apple.com/ipod/carintegration.html), *Apple Inc.*. Retrieved on February 17, 2007.

[41] Apple Teams Up With Continental, Delta, Emirates, & United to deliver iPod Integration (http://www.apple.com/pr/library/2006/nov/14ipod.html), *Apple Inc.*, November 14, 2006. Retrieved on December 7, 2006.

[42] Marsal, Katie. Two of six airlines say there's no ink on iPod deal (http://www.appleinsider.com/article.php?id=2240), *AppleInsider*, November 15, 2006. Retrieved on December 7, 2006.

[43] Machrone, Bill. iPod audio measurements (http://home.comcast.net./~machrone/playertest/playertest.htm), *PC Magazine*, 2005. Retrieved on February 17, 2007.

[44] Heijligers, Marc. iPod audio measurements (http://members.chello.nl/~m.heijligers/ipod/Performance/measurements.html). Retrieved on February 17, 2007.

[45] Heijligers, Marc. iPod circuit design engineering (http://homepage.mac.com/marc.heijligers/audio/ipod/engineering/engineering.html), May 2006. Retrieved on February 17, 2007.

[46] Kuzmanoski, Brian. Tweaking the iPod equalizer (http://www.dapreview.net/p/content/content.php?content.133), *DAP review*, 2005. Retrieved on July 22, 2011.

[47] Cohen, Peter. iPod update limits iPod volume setting (http://www.macworld.com/article/50101/2006/03/ipodupdate.html), Macworld, 2006. Retrieved on November 7, 2008.

[48] Fried, Ian. Apple pulls iPod in France (http://news.cnet.com/2100-1040-960211.html). Retrieved on November 7, 2008.

[49] "iPhone and iPod touch: Charging the battery" (http://support.apple.com/kb/HT1476). Apple. 15 October 2009. . Retrieved 12 January 2010.

[50] Apple Inc., Identifying iPod models (http://support.apple.com/kb/HT1353), retrieved 31 October 2007.

[51] Mactracker ( mactracker.ca (http://www.mactracker.ca/)), Apple Inc. model database, version as of 26 July 2007.

[52] Apple Inc., Apple press release library (http://www.apple.com/pr/library/), Retrieved September 19, 2007.

[53] Apple faces patent lawsuits over its iPod (http://www.channelregister.co.uk/2005/03/10/apple_ipod_patent_lawsuits/), *ChannelRegister*, March 10, 2005. Retrieved on February 17, 2007.

[54] U.S. Patent 6,587,403 (http://www.google.com/patents?vid=USPAT6587403&id=oawMAAAAEBAJ&dq=6,587,403) — Advanced Audio Devices' "music jukebox" patent.

[55] U.S. Patent 6,665,797 (http://www.google.com/patents?vid=USPAT6665797&id=QrR2AAAAEBAJ&dq=6,665,797) — "Protection of software again against unauthorized use" (corrected to "Computer Apparatus/Software Access Control").

[56] Apple, Sony among those named in new DRM lawsuit (http://www.appleinsider.com/article.php?id=1235), *AppleInsider*, August 16, 2005. Retrieved on February 17, 2007.

[57] U.S. patent application 20030095096 (http://appft1.uspto.gov/netacgi/nph-Parser?p=1&u=/netahtml/PTO/search-bool.html&r=1&f=G&l=50&co1=AND&d=PG01&s1="60/346,237"&OS="60/346,237"&RS="60/346,237") Apple Inc.'s application on "rotational user inputs".

[58] U.S. Patent 6,928,433 (http://www.google.com/patents?vid=USPAT6928433&id=sJoVAAAAEBAJ&dq=6,928,433) Creative Technology's "Zen" patent.

[59] Creative wins MP3 player patent (http://news.bbc.co.uk/2/hi/technology/4198360.stm), *BBC News*, August 30, 2005. Retrieved on February 17, 2007.

[60] Creative sues Apple over patent (http://www.macworld.co.uk/news/index.cfm?NewsID=14646), *Macworld UK*, May 16, 2006. Retrieved on March 20, 2007.

[61] Apple & Creative Announce Broad Settlement... (http://www.apple.com/pr/library/2006/aug/23settlement.html), , *Apple Inc.*. Retrieved on February 17, 2007.

[62] Marsal, Katie. iPod: how big can it get? (http://www.appleinsider.com/article.php?id=1770), *AppleInsider*, May 24, 2006. Retrieved on February 17, 2007.

[63] HP to stop selling Apple's iPods (http://www.appleinsider.com/articles/05/07/29/hp_to_stop_selling_apples_ipods.html), *AppleInsider*, July 29, 2005. Retrieved on August 6, 2007.

[64] "Apple Reports First Quarter Results" (http://www.apple.com/pr/library/2007/01/17Apple-Reports-First-Quarter-Results.html). Apple Press Info. January 17, 2007. .

[65] "Apple Reports Second Quarter Results" (http://www.apple.com/pr/library/2007/04/25Apple-Reports-Second-Quarter-Results.html). Apple Press Info. April 25, 2007. .

[66] "Study: iPod users to switch for iPhone" (http://www.macnn.com/articles/07/05/16/ipod.users.back.iphone/). MacNN. May 16, 2007. .

[67] Apple Reports Fourth Quarter 2007 Results (http://www.apple.com/pr/library/2007/10/22results.html), *Apple Inc.*, October 22, 2007. Retrieved on October 22, 2007.

[68] Apple Inc. (January 22, 2008). Apple Reports First Quarter Results (http://www.apple.com/pr/library/2008/01/22results.html). Press release. Retrieved on January 23, 2008

[69] AppleInsider (October 27, 2008). (http://www.appleinsider.com/articles/08/10/27/ipods_accounted_for_just_14_2_of_apples_sales_in_q4_2008.html) Retrieved on October 27, 2008

[70] World of Apple. (September 9, 2009). Live Coverage From Apple's "It's Only Rock and Roll" Event (http://news.worldofapple.com/category/world-of-apple-events/). Press release. Retrieved on September 9, 2009

[71] Albanesius, Chloe (October 4, 2011). "Apple Unveils Updated iPod nano, touch" (http://www.pcmag.com/article2/0,2817,2394061,00.asp). PC Magazine. . Retrieved December 20, 2011.

[72] iPod and Bluetooth lead to prizes (http://news.bbc.co.uk/1/hi/technology/4605881.stm), *BBC News*, June 3, 2005. Retrieved on March 20, 2007.

[73] "The 25 Most Innovative Products of the Year" (http://www.pcworld.com/article/id,123942-page,2/article.html). PC World. August 3, 2007. . Retrieved February 17, 2007.

[74] Apple wins 5 'World Class' awards (http://www.macnn.com/articles/06/05/31/top.100.products.announced/), *MacNN*. Retrieved on February 17, 2007.

[75] Apple, iPod, and CD Copy Protection (http://www.macrumors.com/pages/2004/07/20040720124549.shtml), *MacRumors*. Retrieved on February 17, 2007.

[76] Hospitals train staff with iPods (http://news.bbc.co.uk/1/hi/business/4859302.stm), *BBC News*, March 29, 2006. Retrieved on June 16, 2007.

[77] "Mobile Learning" (http://www.apple.com/education/teachers-professors/mobile-learning.html). Apple.com. . Retrieved 2011-08-15.

[78] Maag, M.E.D. (2006). "Podcasting and MP3 players: emerging education technologies" (http://www.cinjournal.com/pt/re/cin/fulltext.00024665-200601000-00005.htm). *CIN: Computers, Informatics, Nursing* **24** (1): 9–13. PMID 16436906. . Retrieved February 8, 2008.

[79] Slykhuis, D. (2006). *Have an iPod? Then you need to know this about how to use it in your classroom* (http://www.aace.org/newdl/index.cfm?fuseaction=Reader.ViewAbstract&paper_id=22435). . Retrieved February 8, 2008.

[80] — Doug Lederman. "Duke's Ever-Evolving iPod Initiative" (http://www.insidehighered.com/news/2006/04/28/ipod). Insidehighered.com. . Retrieved 2011-08-15.

[81] Geier, Thom; Jensen, Jeff; Jordan, Tina; Lyons, Margaret; Markovitz, Adam; Nashawaty, Chris; Pastorek, Whitney; Rice, Lynette; Rottenberg, Josh; Schwartz, Missy; Slezak, Michael; Snierson, Dan; Stack, Tim; Stroup, Kate; Tucker, Ken; Vary, Adam B.; Vozick-Levinson, Simon; Ward, Kate (December 11, 2009), "THE 100 Greatest MOVIES, TV SHOWS, ALBUMS, BOOKS, CHARACTERS, SCENES, EPISODES, SONGS, DRESSES, MUSIC VIDEOS, AND TRENDS THAT ENTERTAINED US OVER THE PAST 10 YEARS". Entertainment Weekly. (1079/1080):74-84

[82] MP3 Insider: The truth about your battery life (http://www.mp3.com/features/stories/3646.html), *mp3.com*, March 13, 2006. Retrieved on July 10, 2006.

[83] Apple investigates iPod batteries (http://news.bbc.co.uk/1/hi/technology/3477497.stm), *BBC News*, February 10, 2004. Retrieved on March 20, 2007.

[84] Horwitz, Jeremy. Apple's iPod Battery Settlement, Explained (http://www.ilounge.com/index.php/articles/comments/apples-ipod-battery-settlement-explained/), *iLounge*, June 10, 2005. Retrieved on August 27, 2006.

[85] The Curse of Lithium Ion Batteries (http://www.mp3newswire.net/stories/6002/li_ion.html), *MP3 Newswire*, January 6, 2006. Retrieved on November 30, 2006.

[86] iPod Battery FAQ (http://ipodbatteryfaq.com/#18). Retrieved on November 26, 2006.

[87] Neistat, Casey. A Message From the Neistat Brothers (http://www.ipodsdirtysecret.com/message.html), November 20, 2003. Retrieved on February 17, 2007.

[88] Apple offers iPod battery replacement service (http://www.macminute.com/2003/11/14/ipodbattery), *MacMinute*, November 14, 2003. Retrieved on November 26, 2006.

[89] AppleCare for iPod now available (http://www.macminute.com/2003/11/21/ipodapplecare/), *MacMinute*, November 21, 2003. Retrieved on November 26, 2006.

[90] Ecker, Clint. Vivisection of the Video iPod (http://arstechnica.com/reviews/hardware/video-ipod.ars/6), *Ars Technica*, October 19, 2005. Retrieved on November 26, 2006.

[91] Disassemble Guide for Video iPod (http://www.ipodmods.com/guides/Video/videodisassemblyguide.htm). Retrieved on November 26, 2006.

[92] "Apple opens up replacement program for first-gen iPod nano" (http://www.appleinsider.com/articles/11/11/12/apple_opens_up_replacement_program_for_first_gen_ipod_nano.html). Apple Insider. November 12, 2011. . Retrieved December 20, 2011.

[93] iPod Reliability Survey (http://www.macintouch.com/reliability/ipodfailures.html), *MacInTouch*, November 28, 2005. Retrieved on October 29, 2006.

[94] Apple responds to iPod nano screen concerns (http://www.macworld.com/news/2005/09/27/nanoscreen/index.php), *Macworld*, September 27, 2005. Retrieved on February 17, 2007.

[95] Arthur, Charles. iPod nano owners in screen scratch trauma (http://www.theregister.co.uk/2005/09/23/ipod_nano_scratching/), *The Register*, September 25, 2005. Retrieved on February 17, 2007.

[96] Fried, Ina. Suit filed over Nano scratches (http://news.com.com/Suit+filed+over+Nano+scratches/2100-1047_3-5906399.html), *CNet News*, October 21, 2005. Retrieved on February 17, 2007.

[97] Inside Apple's iPod factories (http://www.macworld.co.uk/news/index.cfm?NewsID=14915), *Macworld UK*, June 12, 2006. Retrieved on March 20, 2007.

[98] Millard, Elizabeth. Is It Ethical To Own an iPod? (http://www.chinesetao.com/Newsfactor Sept 2006.pdf). Retrieved on March 20, 2007.

[99] Foxconn denies iPod 'sweatshop' claims (http://www.macnn.com/articles/06/06/19/ipod.sweatshop.claims/), *MacNN*, June 19, 2006. Retrieved on February 17, 2007.

[100] ITUC Hong Kong Liaison Office (January 2, 2007). "ACFTU Union established at Foxconn on the very last day of 2006" (http://www.ihlo.org/LRC/ACFTU/030107.html). . Retrieved February 2, 2010.

[101] McDonald's and KFC seeking to resolve Chinese minimum wage issue ... (http://www.nytimes.com/2007/04/05/business/worldbusiness/05iht-labor.1.5156948.html), April 5, 2007, nytimes.com. Retrieved 2010 5 27.

[102] Wal-Mart backs down and allows Chinese workers to join union (http://www.guardian.co.uk/business/2006/aug/11/china.supermarkets), August 11, 2006, Jonathan Watts, The Guardian

[103] Suicides Spark Inquiries Apple, H-P to Examine Asian Supplier After String of Deaths at Factory (http://online.wsj.com/article/SB10001424052748704026204575267603576594936.html), Jason Dean, Ting-i Tsai, May 27, 2010, accessed May 27, 2010

[104] The Foxconn Suicides (http://online.wsj.com/article/SB10001424052748704269204575270031332376238.html?mod=googlenews_wsj), May 28, 2010, wsj.com, WSJ opinion, accessed May 27, 2010

[105] "Reuters-Freeland-Winners and Losers in the Apple Economy-July 2011" (http://blogs.reuters.com/chrystia-freeland/2011/07/01/winners-and-losers-in-the-apple-economy/). Blogs.reuters.com. July 1, 2011. . Retrieved 2011-08-15.

# External links

- Apple iPod (http://www.apple.com/ipod/)—Official website
- iPod troubleshooting basics and service FAQ (http://docs.info.apple.com/article.html?artnum=32222)—From the official website
- *Apple's 21st century Walkman* (http://money.cnn.com/magazines/fortune/fortune_archive/2001/11/12/313342/index.htm)—Brent Schlender, *Fortune*, October 2001
- iPod Nation (http://www.newsweek.com/id/54529)—Steven Levy, *Newsweek*, July 2004
- The Perfect Thing (http://www.wired.com/wired/archive/14.11/ipod.html)—Steven Levy, *Wired*, November 2006

# Macintosh

The **Macintosh** ( /ˈmækɪntɒʃ/ ***MAK**-in-tosh*),[1] or **Mac**, is a series of personal computers (PC) designed, developed, and marketed by Apple Inc. The first Macintosh was introduced by Apple's then-chairman Steve Jobs on January 24, 1984; it was the first commercially successful personal computer to feature a mouse and a graphical user interface rather than a command-line interface.[2] The company continued to have success through the second half of the 1980s, first primarily because of the sales of the Apple II series remained strong even after the introduction of the Macintosh, only to see it dissipate in the 1990s as the personal computer market shifted toward the "Wintel" platform: IBM PC compatible machines running MS-DOS and Microsoft Windows.[3]

The original Macintosh, the first commercially successful personal computer to use a graphical user interface, rather than a command line.

In 1998, Apple consolidated its multiple consumer-level desktop models into the iMac all-in-one. This proved to be a sales success and saw the Macintosh brand revitalized. Current Mac systems are mainly targeted at the home, education, and creative professional markets. These include the descendants of the original iMac and the entry-level Mac mini desktop models, the Mac Pro tower graphics workstation, the MacBook, MacBook Air and MacBook Pro laptops. The Xserve server was discontinued January 31, 2011.[4]

An iMac computer from October 2009, a modern all-in-one Macintosh.

Production of the Mac is based on a vertical integration model in that Apple facilitates all aspects of its hardware and creates its own operating system that is pre-installed on all Mac computers. This is in contrast to most IBM PC compatibles, where multiple sellers create and integrate hardware intended to run another company's operating software. Apple exclusively produces Mac hardware, choosing internal systems, designs, and prices. Apple does use third party components, however, such as graphics subsystems from nVidia and ATi. Current Mac CPUs use Intel's X86-64 architecture; the earliest models (1984–1994) used Motorola's 68k and models from 1994–2006 used the AIM alliance's PowerPC. Apple also develops the operating system for the Mac, currently Mac OS X version 10.7 "Lion". The modern Mac, like other personal computers, is capable of running alternative operating systems such as Linux, FreeBSD, and, in the case of Intel-based Macs, Microsoft Windows. However, Apple does not license Mac OS X for use on non-Apple computers.

## History

### 1979 to 1984: Development

The Macintosh project started in the late 1970s with Jef Raskin, an Apple employee who envisioned an easy-to-use, low-cost computer for the average consumer. He wanted to name the computer after his favorite type of apple, the McIntosh,[5] but the name had to be changed for legal reasons as it was too close, phonetically, to that of the McIntosh audio equipment manufacturer. Steve Jobs requested a release of the name so that Apple could use it, but was denied, forcing Apple to eventually buy the rights to use the name.[6] Raskin was authorized to start hiring for the project in September 1979,[7] and he began to look for an engineer who could put together a prototype. Bill

Atkinson, a member of Apple's Lisa team (which was developing a similar but higher-end computer), introduced him to Burrell Smith, a service technician who had been hired earlier that year. Over the years, Raskin assembled a large development team that designed and built the original Macintosh hardware and software; besides Raskin, Atkinson and Smith, the team included George Crow,[8] Chris Espinosa, Joanna Hoffman, Bruce Horn, Susan Kare, Andy Hertzfeld, Guy Kawasaki, Daniel Kottke,[9] and Jerry Manock.[10] [11]

Smith's first Macintosh board was built to Raskin's design specifications: it had 64 kilobytes (kB) of RAM, used the Motorola 6809E microprocessor, and was capable of supporting a 256×256 pixel black-and-white bitmap display. Bud Tribble, a Macintosh programmer, was interested in running the Lisa's graphical programs on the Macintosh, and asked Smith whether he could incorporate the Lisa's Motorola 68000 microprocessor into the Mac while still keeping the production cost down. By December 1980, Smith had succeeded in designing a board that not only used the 68000, but increased its speed from 5 MHz to 8 MHz; this board also had the capacity to support a 384×256 pixel display. Smith's design used fewer RAM chips than the Lisa, which made production of the board significantly more cost-efficient. The final Mac design was self-contained and had the complete QuickDraw picture language and interpreter in 64 kB of ROM – far more than most other computers; it had 128 kB of RAM, in the form of sixteen 64 kilobit (kb) RAM chips soldered to the logicboard. Though there were no memory slots, its RAM was expandable to 512 kB by means of soldering sixteen IC sockets to accept 256 kb RAM chips in place of the factory-installed chips. The final product's screen was a 9-inch, 512x342 pixel monochrome display, exceeding the prototypes.[12]

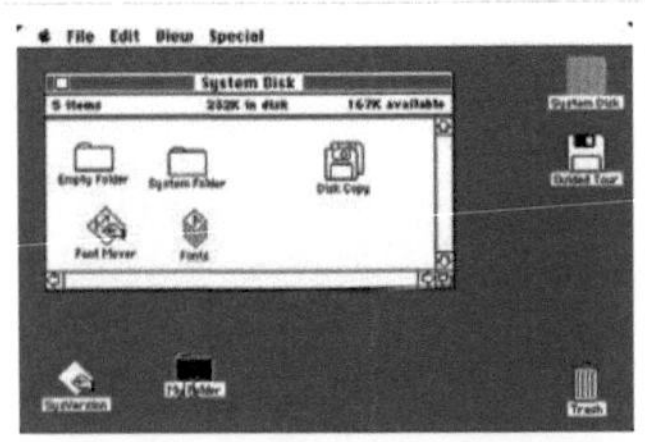

The original 1984 Mac OS desktop featured a radically new graphical user interface. Users communicated with the computer not through abstract textual commands but rather using a metaphorical desktop that included icons of real life items with which the user was already familiar.

The design caught the attention of Steve Jobs, co-founder of Apple. Realizing that the Macintosh was more marketable than the Lisa, he began to focus his attention on the project. Raskin finally left the Macintosh project in 1981 over a personality conflict with Jobs, and team member Andy Hertzfeld said that the final Macintosh design is closer to Jobs' ideas than Raskin's.[7] After hearing of the pioneering GUI technology being developed at Xerox PARC, Jobs had negotiated a visit to see the Xerox Alto computer and Smalltalk development tools in exchange for Apple stock options. The Lisa and Macintosh user interfaces were influenced by technology seen at Xerox PARC and were combined with the Macintosh group's own ideas.[13] Jobs also commissioned industrial designer Hartmut Esslinger to work on the Macintosh line, resulting in the "Snow White" design language; although it came too late for the earliest Macs, it was implemented in most other mid- to late-1980s Apple computers.[14] However, Jobs' leadership at the Macintosh project did not last; after an internal power struggle with new CEO John Sculley, Jobs resigned from Apple in 1985,[15] went on to found NeXT, another computer company,[16] and did not return until 1997 when Apple acquired NeXT.[17]

The first Macintosh computer was manufactured at an Apple plant in Fremont, California.[18]

## 1984: Introduction

This television commercial, first aired during Super Bowl XVIII, launched the original Macintosh.

The Macintosh 128K was announced to the press in October 1983, followed by an 18-page brochure included with various magazines in December.[19] The Macintosh was introduced by the now famous US$1.5 million Ridley Scott television commercial, "1984".[20] The commercial most notably aired during the third quarter of Super Bowl XVIII on January 22, 1984 and is now considered a "watershed event"[21] and a "masterpiece."[22] "1984" used an unnamed heroine to represent the coming of the Macintosh (indicated by a Picasso-style picture of Apple's Macintosh computer on her white tank top) as a means of saving humanity from the "conformity" of IBM's attempts to dominate the computer industry. The ad alludes to George Orwell's novel, *Nineteen Eighty-Four,* which described a dystopian future ruled by a televised "Big Brother."[23] [24]

Two days after the 1984 ad aired, the Macintosh went on sale. It came bundled with two applications designed to show off its interface: MacWrite and MacPaint. It was first demonstrated by Steve Jobs in the first of his famous Mac Keynote speeches, and though the Mac garnered an immediate, enthusiastic following, some labeled it a mere "toy."[25] Because the operating system was designed largely around the GUI, existing text-mode and command-driven applications had to be redesigned and the programming code rewritten. This was a time-consuming task that many software developers chose not to undertake, and could be regarded as a reason for an initial lack of software for the new system. In April 1984 Microsoft's MultiPlan migrated over from MS-DOS, with Microsoft Word following in January 1985.[26] In 1985, Lotus Software introduced Lotus Jazz for the Macintosh platform after the success of Lotus 1-2-3 for the IBM PC, although it was largely a flop.[27] Apple introduced Macintosh Office the same year with the lemmings ad. Infamous for insulting its own potential customers, the ad was not successful.[28]

The Apple Macintosh Plus at the Design Museum in Gothenburg, Sweden.

For a special post-election edition of *Newsweek* in November 1984, Apple spent more than US$2.5 million to buy all 39 of the advertising pages in the issue.[29] Apple also ran a "Test Drive a Macintosh" promotion, in which potential buyers with a credit card could take home a Macintosh for 24 hours and return it to a dealer afterwards. While 200,000 people participated, dealers disliked the promotion, the supply of computers was insufficient for demand, and many were returned in such a bad shape that they could no longer be sold. This marketing campaign caused CEO John Sculley to raise the price from US$1,995 to US$2,495 (adjusting for inflation, about $5,200 in 2010).[28] [30]

## 1985 to 1989: Desktop publishing era

In 1985, the combination of the Mac, Apple's LaserWriter printer, and Mac-specific software like Boston Software's MacPublisher and Aldus PageMaker enabled users to design, preview, and print page layouts complete with text and graphics—an activity to become known as desktop publishing. Initially, desktop publishing was unique to the Macintosh, but eventually became available for other platforms as well.[31] Later, applications such as Macromedia FreeHand, QuarkXPress, Adobe Photoshop, and Adobe Illustrator strengthened the Mac's position as a graphics computer and helped to expand the emerging desktop publishing market.

The limitations of the first Mac soon became clear: it had very little memory, even compared with other personal computers in 1984, and could not be expanded easily; and it lacked a hard disk drive or the means to attach one easily. Many small companies sprang up to address the memory issue, by upgrading the 128k Mac to 512 kB, by removing the computer's 16 memory chips and replacing them with larger-capacity chips, a tedious operation that was not always successful. In October 1985, Apple increased the Mac's memory to 512 kB,[32] and offered an upgrade for 128k Macs that involved replacing the logic board. In an attempt to improve connectivity, Apple released the Macintosh Plus on January 10, 1986 for US$2,600. It offered one megabyte of RAM, easily expandable to four by the use of socketed RAM boards, and a SCSI parallel interface, allowing up to seven peripherals—such as hard drives and scanners—to be attached to the machine. Its floppy drive was increased to an 800 kB capacity. The Mac Plus was an immediate success and remained in production, unchanged, until October 15, 1990; on sale for just over four years and ten months, it was the longest-lived Macintosh in Apple's history.[33]

Updated Motorola CPUs made a faster machine possible, and in 1987 Apple took advantage of the new Motorola technology and introduced the Macintosh II, which used a 16 MHz Motorola 68020 processor.[34] The primary improvement in the Macintosh II was Color QuickDraw in ROM, a color version of the graphics language which was the heart of the machine. Among the many innovations in Color QuickDraw were an ability to handle any display size, any color depth, and multiple monitors.

The Macintosh II, one of the first expandable Macintosh models.

The Macintosh II marked the start of a new direction for the Macintosh, as now for the first time it had an open architecture with several NuBus expansion slots, support for color graphics and external monitors, and a modular design similar to that of the IBM PC. It had an internal hard drive and a power supply with a fan, which was initially fairly loud.[35] One third-party developer sold a device to regulate fan speed based on a heat sensor, but it voided the warranty.[36] Later Macintosh computers had quieter power supplies and hard drives.

In September 1986, Apple introduced the Macintosh Programmer's Workshop, or MPW that allowed software developers to create software for Macintosh on Macintosh, rather than cross compiling from a Lisa. In August 1987, Apple unveiled HyperCard, and introduced MultiFinder, which added cooperative multitasking to the Macintosh. Apple began bundling both with every Macintosh.

The Macintosh SE was released at the same time as the Macintosh II, as the first compact Mac with a 20 MB internal hard drive and one expansion slot.[37] The SE's expansion slot was located inside the case along with the CRT, potentially exposing an upgrader to high voltage. For this reason Apple recommended users bring their SE to an authorized Apple dealer to have upgrades performed.[38] The SE also updated Jerry Manock and Terry Oyama's original design and shared the Macintosh II's *Snow White* design language, as well as the new Apple Desktop Bus (ADB) mouse and keyboard that had first appeared on the Apple IIGS some months earlier.

In 1987, Apple spun off its software business as Claris. It was given the code and rights to several applications that had been written within Apple, notably MacWrite, MacPaint, and MacProject. In the late 1980s, Claris released a number of revamped software titles; the result was the "Pro" series, including MacDraw Pro, MacWrite Pro, and FileMaker Pro. To provide a complete office suite, Claris purchased the rights to the Informix Wingz spreadsheet on the Mac, renaming it Claris Resolve, and added the new presentation software Claris Impact. By the early 1990s, Claris applications were shipping with the majority of consumer-level Macintoshes and were extremely popular. In 1991, Claris released ClarisWorks, which soon became their second best-selling application. When Claris was reincorporated back into Apple in 1998, ClarisWorks was renamed AppleWorks beginning with version 5.0.[39]

In 1988, Apple sued Microsoft and Hewlett-Packard on the grounds that they infringed Apple's copyrighted GUI, citing (among other things) the use of rectangular, overlapping, and resizable windows. After four years, the case was decided against Apple, as were later appeals. Apple's actions were criticized by some in the software community, including the Free Software Foundation (FSF), who felt Apple was trying to monopolize on GUIs in general, and boycotted GNU software for the Macintosh platform for seven years.[40] [41]

The Macintosh Portable was Apple's first battery-powered Macintosh. It was available from 1989 to 1991 and could run System 6 and System 7.

With the new Motorola 68030 processor came the Macintosh IIx in 1988, which had benefited from internal improvements, including an on-board MMU.[42] It was followed in 1989 by a more compact version with fewer slots (the Macintosh IIcx)[43] and a version of the Mac SE powered by the 16 MHz 68030, the Macintosh SE/30.[44] Later that year, the Macintosh IIci, running at 25 MHz, was the first Mac to be "32-bit clean," allowing it to natively support more than 8 MB of RAM,[45] unlike its predecessors, which had "32-bit dirty" ROMs (8 of the 32 bits available for addressing were used for OS-level flags). System 7 was the first Macintosh operating system to support 32-bit addressing.[46] Apple also introduced the Macintosh Portable, a 16 MHz 68000 machine with an active matrix flat panel display that was backlit on some models.[47] The following year the Macintosh IIfx, starting at US$9,900, was unveiled. Apart from its fast 40 MHz 68030 processor, it had significant internal architectural improvements, including faster memory and two Apple II CPUs dedicated to I/O processing.[48]

### 1990 to 1998: Growth and decline

Microsoft Windows 3.0, which began to approach the Macintosh operating system in both performance and feature set, was released in May 1990 and was a less expensive alternative to the Macintosh platform. Apple's response was to introduce a range of relatively inexpensive Macs in October 1990. The Macintosh Classic, essentially a less expensive version of the Macintosh Plus, was the least expensive Mac until early 2001.[49] The 68020-powered Macintosh LC, in its distinctive "pizza box" case, offered color graphics and was accompanied by a new, low-cost 512 × 384 pixel monitor.[50] The Macintosh IIsi was essentially a 20 MHz IIci with only one expansion slot.[51] All three machines sold well,[52] although Apple's profit margin was considerably lower than on earlier machines.[49]

The Macintosh Classic, Apple's early 1990s budget model.

The PowerBook 100 (shown here), 140 and 170 introduced a line of professional laptop Macs. They pioneered notebook ergonomics by placing the keyboard behind a palm rest.

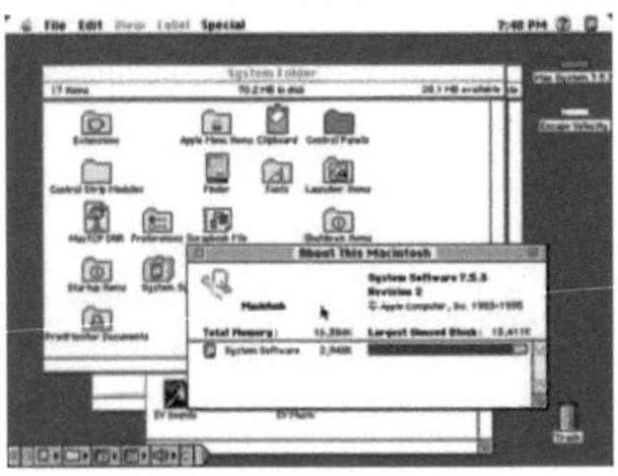

System 7 was the first major upgrade of the Macintosh operating system.

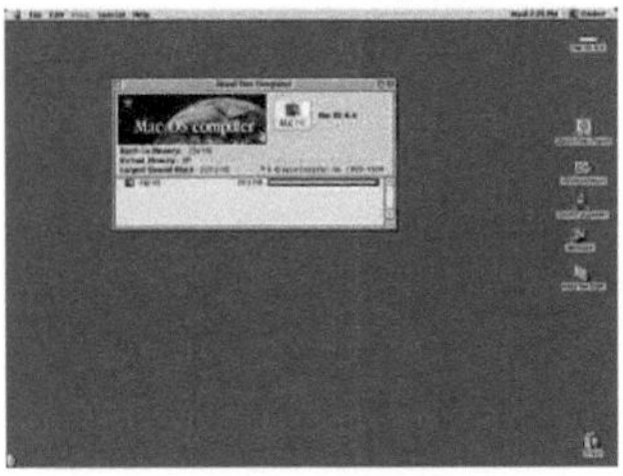

OS 8 was the second major upgrade of the Mac OS. OS 8.6 shown.

Apple improved Macintosh computers by introducing models equipped with newly available processors from the 68k lineup. The Macintosh Classic II[53] and Macintosh LC II, which used a 16 MHz 68030 CPU,[54] were joined in 1991 by the Macintosh Quadra 700[55] and 900,[56] the first Macs to employ the faster Motorola 68040 processor. In 1994, Apple abandoned Motorola CPUs for the RISC PowerPC architecture developed by the AIM alliance of Apple Computer, IBM, and Motorola.[57] The Power Macintosh line, the first to use the new chips, proved to be highly successful, with over a million PowerPC units sold in nine months.[58]

Apple replaced the Macintosh Portable in 1991 with the first of the PowerBook line: the PowerBook 100, a miniaturized Portable; the 16 MHz 68030 PowerBook 140; and the 25 MHz 68030 PowerBook 170.[59] They were the first portable computers with the keyboard behind a palm rest, and with a built-in pointing device (a trackball) in front of the keyboard.[60] The 1993 PowerBook 165c was Apple's first portable computer to feature a color screen, displaying 256 colors with 640 x 400 pixel resolution.[61] The second-generation of PowerBooks, the 68040-equipped 500 series, introduced the trackpad, integrated stereo speakers and built-in Ethernet to the laptop form factor in 1994.[62]

As for Mac OS, System 7 was a 32-bit rewrite from Pascal to C++ that introduced virtual memory, and improved the handling of color graphics, memory addressing, networking, and co-operative multitasking. Also during this time, the Macintosh began to shed the "Snow White" design language, along with the expensive consulting fees they were paying to Frogdesign, in favor of bringing the work in-house by establishing the Apple Industrial Design Group. They became responsible for to crafting a new look to go with the new operating system and all other Apple products.[63]

Despite these technical and commercial successes, Microsoft and Intel began to rapidly lower Apple's market share with the Windows 95 operating system and Pentium processors respectively. These significantly enhanced the multimedia capability and performance of IBM PC compatible computers, and brought Windows still closer to the

Mac GUI. Furthermore, Apple had created too many similar models that confused potential buyers. At one point Apple's product lineup was subdivided into Classic, LC, II, Quadra, Performa, and Centris models, with essentially the same computer being sold under a number of different names.[64] These models competed against the Macintosh clones, hardware manufactured by third-parties that ran Apple's System 7. This succeeded in increasing the Macintosh's market share somewhat and provided cheaper hardware for consumers, but hurt Apple financially as existing Apple customers began to buy cheaper clones while Apple shouldered the burden of developing the platform.

When Steve Jobs returned to Apple in 1997, he ordered that the OS that had been previewed as version 7.7 be branded Mac OS 8 (in place of the never-to-appear Copland OS). Since Apple had licensed only System 7 to third-parties, this move effectively ended the clone line. The decision caused significant financial losses for companies like Motorola, who produced the StarMax, Umax, who produced the SuperMac,[65] and Power Computing Corporation, who offered several lines Mac clones, including PowerWave, PowerTower, and PowerTower Pro.[66] These companies had invested substantial resources in creating their own Mac-compatible hardware.[67] Apple bought out Power Computing's license, but allowed Umax to continue selling Mac clones until their license expired, as they had a sizeable presence in the lower-end segment that Apple did not.

## 1998 to 2005: New beginnings

In 1998, a year after Steve Jobs had returned to the company, Apple introduced an all-in-one Macintosh called the iMac. Its translucent plastic case, originally Bondi blue and later many other colors, is considered an industrial design landmark of the late 1990s. The iMac did away with most of Apple's standard (and usually proprietary) connections, such as SCSI and ADB, in favor of two USB ports. It also had no internal floppy disk drive and instead included a CD-ROM drive for installing software,[3] [69] but was incapable of writing to CDs or other media without external third-party hardware. The iMac proved to be phenomenally successful, with 800,000 units sold in 139 days,[70] making the company an annual profit of US$309 million—Apple's first profitable year since Michael Spindler took over as CEO in 1995.[71] The iMacs's "blue and white" aesthetic was applied to the Power Macintosh, and then to a new product, the iBook. Introduced in July 1999, the iBook was Apple's first consumer-level laptop computer, filling the missing quadrant of Apple's "four-square product matrix" (*desktop* and *portable* products for both *consumers* and *professionals*).[72] More than 140,000 pre-orders were placed before it started shipping in September,[73] and by October it was as much a sales hit as the iMac.[74]

The original "Bondi Blue" iMac G3. Introduced in 1998, it led Apple's return to profitability. However, the associated mouse proved to be one of consumers' least favorite Apple products.[68]

In early 2001, Apple began shipping computers with CDRW drives for the first time.[75] Apple had been emphasizing the Mac's ability to play DVDs by including DVD-ROM and DVD-RAM drives as standard. Steve Jobs admitted that Apple had been "late to the party" on writable CD technology but felt that Macs could become a "digital hub" that linked and enabled an "emerging digital lifestyle".[76] Apple would later introduce an update to its iTunes music player software that could burn CDs, along with a controversial "Rip, Mix, Burn" advertising campaign that some[77] felt encouraged media piracy.[78] This accompanied the release of the iPod, Apple's first successful handheld device.

Apple continued to add new products to their lineup, such as the Power Mac G4 Cube,[79] the eMac for the education market and PowerBook G4 laptop for professionals. The original iMac used a G3 processor, but the G4 and then G5 chips were accompanied by successive new designs, dropping the array of colors in favor of white plastic. As of

2011 iMacs used aluminum enclosures. On January 11, 2005, Apple announced the release of the Mac Mini priced at US$499,[80] the least expensive Mac at its release.[81]

Mac OS continued to evolve up to version 9.2.2, including retrofits such as the addition of a nanokernel and support for Multiprocessing Services 2.0 in Mac OS 8.6.[82] Ultimately its dated architecture made replacement necessary. Initially developed in the Pascal programming language, it was substantially rewritten in C++ for System 7. From its beginnings on a 128k 8 MHz machine, it had grown to support Apple's latest 1 GHz G4-equipped Macs. But since its architecture was laid down, OS features like preemptive multitasking and protected memory had become feasible on the kind of hardware Apple manufactured – features that were already common on Apple's competition. As such, Apple introduced Mac OS X, a fully overhauled Unix-based successor to Mac OS 9, using Darwin, XNU, and Mach as foundations, and based on NEXTSTEP. Mac OS X was not released to the public until September 2000, as the Mac OS X Public Beta, with a revamped user interface Apple called "Aqua". At US$29.99, it allowed adventurous Mac users to sample Apple's new operating system and provide feedback for the actual release.[83] The initial release of Mac OS X, 10.0 (nicknamed Cheetah), was released on March 24, 2001. Older Mac OS applications could still run under early Mac OS X versions, using an environment called Classic. Subsequent releases of Mac OS X were 10.1 "Puma" (September 25, 2001), 10.2 "Jaguar" (August 24, 2002), 10.3 "Panther" (October 24, 2003) and 10.4 "Tiger" (April 29, 2005).

## 2006 onward: Intel era

The MacBook Pro is the first Mac notebook to use an Intel processor. It was released at Macworld 2006.[84]

Apple discontinued the use of PowerPC microprocessors in 2006. At WWDC 2005, Steve Jobs revealed this transition and also noted that Mac OS X was in development to run both on Intel and PowerPC architecture from the very beginning.[85] All new Macs now use x86 processors made by Intel, and some Macs were given new names to signify the switch.[86] Intel-based Macs can run pre-existing software developed for PowerPC using an emulator called Rosetta,[87] although at noticeably slower speeds than native programs, and the Classic environment is unavailable. Intel chips introduced the potential to run the Microsoft Windows operating system natively on Apple hardware, without emulation software such as Virtual PC. In March 2006, a group of hackers announced that they were able to run Windows XP on an Intel-based Mac. The group released their software as open source and has posted it for download on their website.[88] On April 5, 2006, Apple announced the public beta availability of their own Boot Camp software that allows owners of Intel-based Macs to install Windows XP on their machines; later versions added support for Windows Vista and Windows 7. Boot Camp became a standard feature in Mac OS X 10.5, while support for Classic was dropped from PowerPC Macs.[89] [90]

Starting in 2006, Apple's industrial design has shifted to favor aluminum, which was used in the construction of the first MacBook Pro. Glass was added in 2008 with the introduction of the unibody MacBook Pro. These materials are billed as environmentally friendly.[91] The iMac, MacBook Pro and MacBook Air lines use aluminum enclosures, and are now made of a single unibody.[92] [93] Chief designer Jonathan Ive continues to guide products towards a minimalist and simple feel,[94] [95] including the elimination of replaceable batteries in notebooks.[96] Multi-touch gestures from the iPhone's interface have been applied to the Mac line in the form of touch pads on notebooks and the Magic Mouse and Magic Trackpad for desktops.

In recent years, Apple has seen a significant boost in sales of Macs. This has been attributed, in part, to the success of the iPod and the iPhone, a halo effect whereby satisfied iPod or iPhone owners purchase more Apple equipment,[97] as well as the use of Intel microprocessors.[98] From 2001 to 2008, Mac sales increased continuously on an annual basis. Apple reported worldwide sales of 3.36 million Macs during the 2009 holiday season.[99] As of Mid-2011, Macintosh continues to enjoy rapid market share increase in the US, growing from 7.3% of all computer shipments in 2010 to 9.3% in 2011.[100]

On February 24, 2011, Apple was the first company to bring to market a computer that utilized Intel's new Thunderbolt (codename Light Peak) I/O interface. Using the same physical interface as a Mini DisplayPort, and backwards compatible with that standard, Thunderbolt boasts two-way transfer speeds of 10 Gbit/s.[101]

## Product line

| | Compact | Consumer | Professional |
|---|---|---|---|
| **Desktop** | **Mac mini** 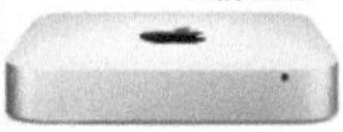 Entry-level; ships without keyboard, mouse, or monitor; uses Intel Core i5, or Intel Core i7 processors | **iMac**  All-in-one; available in 21.5" and 27" screen sizes; uses Intel Core i3, Intel Core i5, or Intel Core i7 processors | **Mac Pro**  Workstation desktop; highly customizable; Available with 4, 8 or 12 cores. Uses up-to two Intel Xeon X5650 "Westmere" (hexa-core) or Xeon X3460 "Nehalem" quad-core processors |
| **Portable (MacBook)** | **MacBook Air 11**  11.6" ultraportable with aluminum casing; uses Intel Core i5, or Intel Core i7 processors | **MacBook Air 13**  13.3" ultraportable with aluminum casing; uses Intel Core i5, or Intel Core i7 processors | **MacBook Pro**  13.3", 15.4" or 17" models with aluminum casing; uses Intel Core i5, or Intel Core i7 processors |
| **Server** | **Mac mini Server**  An additional Mac mini configuration that ships with Mac OS X Server installed. | | **Mac Pro Server**  An additional Mac Pro server configuration that ships with Mac OS X Server installed. |

## Hardware and software

### Hardware

Apple directly sub-contracts hardware production to Asian original equipment manufacturers such as Asus, maintaining a high degree of control over the end product. By contrast, most other companies (including Microsoft) create software that can be run on hardware produced by a variety of third-parties, like Dell, HP/Compaq, and Lenovo. Consequently, the Macintosh buyer has comparably fewer options.

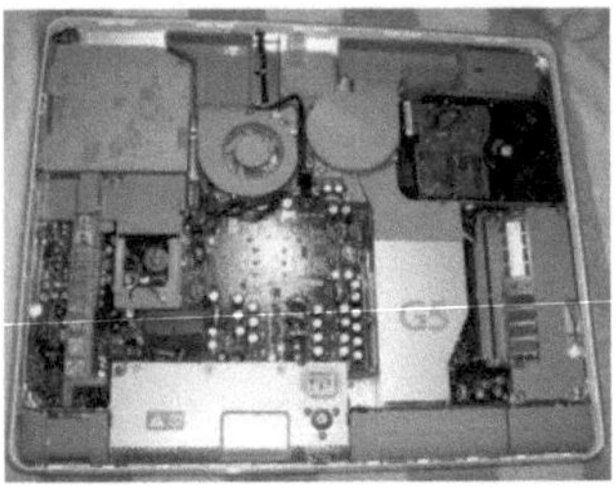
The internals of the original 20-inch iMac G5.

The current Mac product family uses Intel x86-64 processors. Apple introduced an emulator during the transition from PowerPC chips (called Rosetta), much as it did during the transition from Motorola 68000 architecture a decade earlier. The Macintosh is the only mainstream computer platform to have successfully transitioned to a new CPU architecture,[102] and has done so twice. All current Mac models ship with at least 2 GB RAM as standard. Current Mac computers use ATI Radeon or nVidia GeForce graphics cards as well as Intel graphics built into the main CPU. All current Macs (except for the MacBook Air and Mac Mini) ship with an optical media drive that includes a dual-function DVD/CD burner, called the SuperDrive. Macs include two standard data transfer ports: USB and FireWire (except for the MacBook Air and MacBook that do not include FireWire). MacBook Pro, iMac, and MacBook Air computers now also feature the "Thunderbolt" port, which Apple says can transfer data at speeds up to 10 gigabits per second. USB was introduced in the 1998 iMac G3 and is ubiquitous today,[3] while FireWire is mainly reserved for high-performance devices such as hard drives or video cameras. Starting with a new iMac G5 released in October 2005, Apple started to include built-in iSight cameras to appropriate models, and a media center interface called Front Row that can be operated by an Apple Remote or keyboard for accessing media stored on the computer.[103]

Apple was initially reluctant to embrace mice with multiple buttons and scroll wheels. Macs did not natively support multiple buttons, even from third parties, until Mac OS X arrived in 2001.[104] Apple continued to offer only single button mice, with wired and Bluetooth wireless versions, until August 2005, when it introduced the Mighty Mouse. While it looked like a traditional one-button mouse, it actually had four buttons and a scroll ball, capable of independent *x*- and *y*-axis movement.[105] A Bluetooth version followed in July 2006.[106] In October 2009, Apple introduced the Magic Mouse which uses multi-touch gesture recognition similar to the iPhone instead of a physical scroll wheel or ball.[107] It is available only in Bluetooth, and the Mighty Mouse (re-branded as "Apple Mouse") is available with a cord. Apple also features the "Magic Trackpad" as a means to control Macintosh desktop computers such as the iMac or Mac Pro. This was introduced in 2010.

### Software

The original Macintosh was the first successful personal computer to use a graphical user interface devoid of a command line. It used a desktop metaphor, depicting real-world objects like documents and a trashcan as icons onscreen. The *System* software introduced in 1984 with the first Macintosh and renamed *Mac OS* in 1997, continued to evolve until version 9.2.2. In 2001, Apple introduced Mac OS X, based on Darwin and NEXTSTEP; its new features included the Dock and the Aqua user interface. During the transition, Apple included an emulator known as Classic allowing users to run Mac OS 9 applications under Mac OS X, version 10.4 and earlier on PowerPC machines. The most recent version is Mac OS X v10.7 "Lion." In addition to Lion, all new Macs are bundled with assorted Apple-produced applications, including iLife, the Safari web browser and the iTunes media player. Apple released Mac OS X 10.7 in 2010, which was made available in the summer of 2011. This operating system includes many new features such as: Mission Control, the Mac App Store (available to Mac OS X v10.6.6 "Snow Leopard."

users by software update),Launchpad, an application viewer and launcher akin to the iOS Home Screen, and Resume, a feature similar to the hibernate function found in Microsoft Windows.

Historically Mac OS X enjoyed a near-absence of the types of malware and spyware that affect Microsoft Windows users.[108] [109] [110] Mac OS X has a smaller usage share compared to Microsoft Windows (roughly 5% and 92%, respectively),[111] but it also has secure UNIX roots. Worms as well as potential vulnerabilities were noted in February 2006, that led some industry analysts and anti-virus companies to issue warnings that Apple's Mac OS X is not immune to malware.[112] Increasing market share coincided with additional reports of a variety of attacks.[113] Apple releases security updates for its software.[114] In early 2011 Mac OS X experienced a large increase in malware attacks,[115] and malware such as Mac Defender, MacProtector, MacGuard were seen as an increasing problem for Mac users. At first, the malware installer required the user to enter the administrative password but later versions were able to install without user input[116] Initially Apple support staff were instructed not to assist in the removal of the malware or admit the existence of the malware issue, but as the malware spread a support document was issued and Apple announced a OS X update to fix the problem. An estimated 100,000 users were affected.[117] [118]

Originally, the hardware architecture was so closely tied to the Mac OS operating system that it was impossible to boot an alternative operating system. The most common workaround, used even by Apple for A/UX, was to boot into Mac OS and then to hand over control to a program that took over the system and acted as a boot loader. This technique was no longer necessary with the introduction of Open Firmware-based PCI Macs, though it was formerly used for convenience on many Old World ROM systems due to bugs in the firmware implementation. Now, Mac hardware boots directly from Open Firmware (most PowerPC-based Macs) or EFI (all Intel-based Macs), and Macs are no longer limited to running just Mac OS X.

Following the release of the Intel-based Mac, third-party platform virtualization software such as Parallels Desktop, VMware Fusion, and VirtualBox began to emerge. These programs allow users to run Microsoft Windows or previously Windows-only software on Macs at near native speed. Apple also released Boot Camp and Mac-specific Windows drivers that help users to install Windows XP or Vista and natively dual boot between Mac OS X and Windows. Though not condoned by Apple, it is possible to run the Linux operating system using Boot camp or other virtualization workarounds.[119] [120]

Because Mac OS X is a UNIX-like operating system, borrowing heavily from FreeBSD, many applications written for Linux or BSD run on Mac OS X, often using X11. Apple's smaller market share than Microsoft's means that a smaller range of shareware is available, but many popular commercial software applications from large developers such as Microsoft Office and Adobe Photoshop are ported to both Mac OS and Windows. And much of open source software like the Firefox web browser and the OpenOffice.org office suite are cross-platform and run natively. But Mac OS X does not have as many games and "Just for fun" apps as Windows does, so some still prefer Windows.

## Advertising

Apple hyped the introduction of the original Mac with their 1984 commercial that aired during the Super Bowl.[121] It was supplemented by a number of printed pamphlets and other TV ads demonstrating the new interface and emphasizing the mouse. Many more brochures for new models like the Macintosh Plus and the Performa followed. In the 1990s, Apple started the "What's on your PowerBook?" campaign, with print ads and television commercials featuring celebrities describing how the PowerBook helps them in their businesses and everyday lives. In 1995, Apple responded to the introduction of Windows 95 with several print ads and a television commercial demonstrating its disadvantages and lack of innovation. In 1997 the Think Different campaign introduced Apple's new slogan, and in 2002 the Switch campaign followed. The most recent advertising strategy by Apple is the Get a Mac campaign, with North American, UK and Japanese variants.[122] [123]

Today, Apple introduces new products at "special events" at the Apple Town Hall auditorium, and keynotes at the Apple Worldwide Developers Conference, and (formerly) trade shows like the Apple Expo and the MacWorld Expo.

The events typically draw a large gathering of media representatives and spectators, and are preceded by speculation about possible new products. In the past, special events have been used to unveil its desktop and notebook computers such as the iMac and MacBook, and other consumer electronic devices like the iPod, Apple TV, and iPhone, as well as provide updates on sales and market share statistics. Apple has begun to focus its advertising on its retail stores instead of these trade shows; the last MacWorld keynote was in 2009.[124]

## Market share and user demographics

Since the introduction of the Macintosh, Apple has struggled to gain a significant share of the personal computer market. At first, the Macintosh 128K suffered from a dearth of available software compared to IBM's PC, resulting in disappointing sales in 1984 and 1985. It took 74 days for 50,000 units to sell.[125] Market share is measured by browser hits, sales and installed base. If using the browser metric, Mac market share has increased substantially in 2007.[126] If measuring market share by installed base, there were more than 20 million Mac users by 1997, compared to an installed base of around 340 million Windows PCs.[127] [128] Statistics from late 2003 indicate that Apple had 2.06 percent of the desktop share in the United States that had increased to 2.88 percent by Q4 2004.[129] As of October 2006, research firms IDC and Gartner reported that Apple's market share in the U.S. had increased to about 6 percent.[130] Figures from December 2006, showing a market share around 6 percent (IDC) and 6.1 percent (Gartner) are based on a more than 30 percent increase in unit sale from 2005 to 2006. The installed base of Mac computers is hard to determine, with numbers ranging from 5% (estimated in 2009)[131] to 16% (estimated in 2005).[132] Mac OS X's share of the OS market increased from 7.31% in December 2007 to 9.63% in December 2008, which is a 32% increase in market share during 2008, compared with a 22% increase during 2007.

By March 2011, OS X market share in North America had increased to slightly over 14%.[133] Whether the size of the Mac's market share and installed base is relevant, and to whom, is a hotly debated issue. Industry pundits have often called attention to the Mac's relatively small market share to predict Apple's impending doom, particularly in the early and mid 1990s when the company's future seemed bleakest. Others argue that market share is the wrong way to judge the Mac's success. Apple has positioned the Mac as a higher-end personal computer, and so it may be misleading to compare it to a budget PC.[134] Because the overall market for personal computers has grown rapidly, the Mac's increasing sales numbers are effectively swamped by the industry's expanding sales volume as a whole. Apple's small market share, then, gives the impression that fewer people are using Macs than did ten years ago, when exactly the opposite is true.[135] Soaring sales of the iPhone and iPad mean that the portion of Apple's profits represented by the Macintosh has declined in 2010, dropping to 24% from 46% two years earlier.[136] Others try to de-emphasize market share, citing that it is rarely brought up in other industries.[137] Regardless of the Mac's market share, Apple has remained profitable since Steve Jobs' return and the company's subsequent reorganization.[138] Notably, a report published in the first quarter of 2008 found that Apple had a 14% market share in the personal computer market in the US, including 66% of all computers over $1,000.[139] Market research indicates that Apple draws its customer base from a higher-income demographic than the mainstream personal computer market.[140]

## See also

- Apple Inc. litigation
- Apple community
- History of computing hardware (1960s-present)
- List of Macintosh models by case type
- List of Macintosh models grouped by CPU type
- List of Macintosh software
- List of Macintosh software published by Microsoft
- Macintosh User Groups
- Mac gaming

- Reality distortion field
- Lilith (computer)

## Notes


[1] "Define Macintosh" (http://dictionary.reference.com/browse/macintosh). Dictionary.com. . Retrieved April 11, 2010.

[2] Polsson, Ken (July 29, 2009). "Chronology of Apple Computer Personal Computers" (http://www.islandnet.com/~kpolsson/applehis/appl1984.htm). . Retrieved August 27, 2009. See May 3, 1984.

[3] Edwards, Benj (August 15, 2008). "Eight ways the iMac changed computing" (http://www.macworld.com/article/135017/2008/08/imacanniversary.html). Macworld. . Retrieved August 27, 2009.

[4] "Xserve Transition Guide" (http://images.apple.com/xserve/pdf/L422277A_Xserve_Guide.pdf). November 5, 2010. . Retrieved November 5, 2010.

[5] Raskin, Jef (1996). "Recollections of the Macintosh project" (http://mxmora.best.vwh.net/JefRaskin.html). *Articles from Jef Raskin about the history of the Macintosh.*. . Retrieved November 27, 2008.

[6] Apple confidential 2.0: the definitive history of the world's most colorful company, Owen W. Linzmayer, ISBN 978-1-59327-010-0

[7] Hertzfeld, Andy. "The father of the Macintosh" (http://www.folklore.org/StoryView.py?project=Macintosh&story=The_Father_of_The_Macintosh.txt&showcomments=1#comments). Folklore.org. . Retrieved April 24, 2006.

[8] Crow, George. "The Original Macintosh" (http://www.folklore.org/ProjectView.py?project=Macintosh&characters=George Crow). Folklore.org. . Retrieved April 28, 2010.

[9] Kottke, Dan. "The Original Macintosh" (http://www.folklore.org/ProjectView.py?project=Macintosh&characters=Dan Kottke). Folklore.org. . Retrieved April 28, 2010.

[10] Manock, Jerry. "The Original Macintosh" (http://web.archive.org/web/20070929111006/http://folklore.org/ProjectView.py?project=Macintosh&characters=Jerry Manock). Folklore.org. Archived from the original (http://folklore.org/ProjectView.py?project=Macintosh&characters=Jerry Manock) on September 29, 2007. . Retrieved April 28, 2010.

[11] Kawasaki, Guy (January 26, 2009). "Macintosh 25th Anniversary Reunion: Where Did Time Go?" (http://blog.guykawasaki.com/2009/01/twenty-five-yea.html#axzz0mQb8tIS0). . Retrieved April 28, 2010.

[12] Hertzfeld, Andy. "Five different Macintoshes" (http://www.folklore.org/StoryView.py?story=Five_Different_Macs.txt). Folklore.org. . Retrieved April 24, 2006.

[13] Horn, Bruce. "On Xerox, Apple and Progress" (http://www.folklore.org/StoryView.py?project=Macintosh&story=On_Xerox,_Apple_and_Progress.txt). Folklore.org. . Retrieved February 3, 2007.

[14] Tracy, Ed. "History of computer design: Snow White" (http://www.landsnail.com/apple/local/design/design2.html). Landsnail.com. . Retrieved April 24, 2006.

[15] Hertzfeld, Andy. "The End Of An Era" (http://www.folklore.org/StoryView.py?project=Macintosh&story=The_End_Of_An_Era.txt). *folklore.org*. .

[16] Spector, G (September 24, 1985). "Apple's Jobs Starts New Firm, Targets Education Market". *PCWeek*: p. 109.

[17] "Apple Computer, Inc. Finalizes Acquisition of NeXT Software Inc." (http://web.archive.org/web/19990117075346/http://product.info.apple.com/pr/press.releases/1997/q2/970207.pr.rel.next.html). Apple. February 7, 1997. Archived from the original (http://product.info.apple.com/pr/press.releases/1997/q2/970207.pr.rel.next.html) on January 17, 1999. . Retrieved April 27, 2010.

[18] Dennis, Rob (2011-12-30). "Fremont mayor Bob Wasserman dead at 77" (http://www.insidebayarea.com/news/ci_19641994). *The Argus (Fremont)*. . Retrieved 2012-01-21.

[19] "Apple Macintosh 18 Page Brochure" (http://www.digibarn.com/collections/ads/apple-mac/index.htm). DigiBarn Computer Museum. . Retrieved April 24, 2006.

[20] Linzmayer, Owen W. (2004). *Apple Confidential 2.0* (http://www.owenink.com). No Starch Press. p. 113. ISBN 1-59327-010-0. .

[21] Maney, Kevin (January 28, 2004). "Apple's '1984' Super Bowl commercial still stands as watershed event" (http://www.usatoday.com/tech/columnist/kevinmaney/2004-01-28-maney_x.htm). *USA Today*. . Retrieved April 11, 2010.

[22] Leopold, Todd (February 3, 2006). "Why 2006 isn't like '1984'" (http://edition.cnn.com/2006/SHOWBIZ/02/02/eye.ent.commercials/). CNN. . Retrieved May 10, 2008.

[23] Cellini, Adelia (January 2004). "The Story Behind Apple's '1984' TV commercial: Big Brother at 20" (http://web.archive.org/web/20080626112220/http://findarticles.com/p/articles/mi_hb197/is_200401/ai_n5556112). *MacWorld 21.1, page 18*. Archived from the original (http://findarticles.com/p/articles/mi_hb197/is_200401/ai_n5556112) on June 26, 2008. . Retrieved May 9, 2008.

[24] Long, Tony (January 22, 2007). "Jan. 22, 1984: Dawn of the Mac" (http://www.wired.com/science/discoveries/news/2007/01/72496). Wired. . Retrieved April 11, 2010.

[25] Kahney, Leander (January 6, 2004). "We're All Mac Users Now" (http://www.wired.com/gadgets/mac/news/2004/01/61730). Wired. . Retrieved April 11, 2010.

[26] Polsson, Ken. "Chronology of Apple Computer Personal Computers" (http://www.islandnet.com/~kpolsson/applehis/appl1984.htm). . Retrieved November 18, 2007.

[27] Beamer, Scott (January 13, 1992). "For Lotus, third time's the charm" (http://www.accessmylibrary.com/article-1G1-11721498/lotus-third-time-charm.html). MacWEEK. . Retrieved June 23, 2010.

[28] Hormby, Thomas (October 2, 2006). "Apple's Worst Business Decisions" (http://www.osnews.com/story/16036/Apples-Worst-Business-Decisions/). OS News. . Retrieved December 24, 2007.

[29] "1984 *Newsweek* Macintosh ads" (http://www.guidebookgallery.org/ads/magazines/macos/macos10-newsweek). GUIdebook, Newsweek. . Retrieved April 24, 2006.

[30] "Inflation Calculator" (http://data.bls.gov/cgi-bin/cpicalc.pl). Bureau of Labor Statistics. . Retrieved May 14, 2010.

[31] Spring, Michael B. (1991). *Electronic printing and publishing: the document processing revolution* (http://books.google.com/?id=_MV46vFUrI4C&pg=PA125&dq=Desktop+publishing+history+of&cd=5#v=onepage&q=Desktop publishing history of). CRC Press. pp. 125–126. ISBN 9780824785444. .

[32] Technical specifications of Macintosh 512K from Apple's knowledge base (http://support.apple.com/kb/SP187) and from EveryMac.com (http://www.everymac.com/systems/apple/mac_classic/stats/mac_512k.html). Retrieved June 23, 2010.

[33] Technical specifications of Macintosh Plus from Apple's knowledge base (http://support.apple.com/kb/SP190) and from EveryMac.com (http://www.everymac.com/systems/apple/mac_classic/stats/mac_plus.html). Retrieved June 23, 2010.

[34] Technical specifications of Macintosh II from Apple's knowledge base (http://support.apple.com/kb/SP193) and from EveryMac.com (http://www.everymac.com/systems/apple/mac_ii/stats/mac_ii.html). Retrieved June 23, 2010.

[35] "Apple Macintosh II" (http://www.old-computers.com/museum/computer.asp?st=1&c=160). Old Computers On-line Museum. . Retrieved December 23, 2007.

[36] "Macintosh II Family: Fan Regulator Voids Warranty" (http://support.apple.com/kb/TA40560). Apple. July 2, 1992. . Retrieved December 23, 2007.

[37] Technical specifications of Macintosh SE from Apple's knowledge base (http://support.apple.com/kb/SP191) and from EveryMac.com (http://www.everymac.com/systems/apple/mac_classic/stats/mac_se.html). Retrieved June 23, 2010.

[38] "Apple Service Source: Apple SE/30" (http://www.apple-collection.com/CarPos/macse30.pdf). . Retrieved November 16, 2010.

[39] Hearm, Bob (2003). "A Brief History of ClarisWorks" (http://www-swiss.ai.mit.edu/~bob/clarisworks.php). MIT Project on Mathematics and Computation. . Retrieved December 24, 2007.

[40] Free Software Foundation (June 11, 1988). "Special Report: Apple's New Look and Feel" (http://www.gnu.org/bulletins/bull5.html#SEC9). *GNU's Bulletin* **1** (5). . Retrieved April 25, 2006.

[41] Free Software Foundation (1995-01). "End of Apple Boycott" (http://www.gnu.org/bulletins/bull18.html#SEC13). *GNU's Bulletin* **1** (18). . Retrieved April 25, 2006.

[42] Technical specifications of Macintosh IIx from Apple's knowledge base (http://support.apple.com/kb/SP194) and from EveryMac.com (http://www.everymac.com/systems/apple/mac_ii/stats/mac_iix.html). Retrieved June 23, 2010.

[43] Technical specifications of Macintosh IIcx from Apple's knowledge base (http://support.apple.com/kb/{{{appleNumber}}}) and from EveryMac.com (http://www.everymac.com/systems/apple/apple/mac_ii/stats/mac_iicx.html). Retrieved June 23, 2010.

[44] Technical specifications of Macintosh SE/30 from Apple's knowledge base (http://support.apple.com/kb/SP195) and from EveryMac.com (http://www.everymac.com/systems/apple/mac_classic/stats/mac_se30.html). Retrieved June 23, 2010.

[45] Technical specifications of Macintosh IIci from Apple's knowledge base (http://support.apple.com/kb/SP197) and from EveryMac.com (http://www.everymac.com/systems/apple/mac_ii/stats/mac_iici.html). Retrieved June 23, 2010.

[46] Knight, Dan (2001-01). "32-bit Addressing on Older Macs" (http://lowendmac.com/trouble/32bit.shtml). Low End Mac. . Retrieved December 24, 2007.

[47] Technical specifications of Macintosh Portable from Apple's knowledge base (http://support.apple.com/kb/SP140) and from EveryMac.com (http://www.everymac.com/systems/apple/powerbook/stats/mac_portable.html). Retrieved June 23, 2010.

[48] Technical specifications of Macintosh IIfx from Apple's knowledge base (http://support.apple.com/kb/SP203) and from EveryMac.com (http://www.everymac.com/systems/apple/mac_ii/stats/mac_iifx.html). Retrieved June 23, 2010.

[49] Fisher, Lawrence M. (October 15, 1990). "Less-Costly Apple Line To Be Presented Today" (http://query.nytimes.com/gst/fullpage.html?res=9C0CEFD9163CF936A25753C1A966958260). *The New York Times*. . Retrieved January 16, 2008.

[50] Technical specifications of Macintosh LC from Apple's knowledge base (http://support.apple.com/kb/SP205) and from EveryMac.com (http://www.everymac.com/systems/apple/mac_lc/stats/mac_lc.html). Retrieved June 24, 2010.

[51] Technical specifications of Macintosh IIsi from Apple's knowledge base (http://support.apple.com/kb/SP199) and from EveryMac.com (http://www.everymac.com/systems/apple/mac_ii/stats/mac_iisi.html). Retrieved June 24, 2010.

[52] Fisher, Lawrence M. (January 18, 1991). "I.B.M. Surprises Wall Street With Strong Quarterly Net; Apple Posts 20.6% Rise" (http://query.nytimes.com/gst/fullpage.html?res=9D0CE2DA163BF93BA25752C0A967958260). *The New York Times*. . Retrieved January 16, 2008.

[53] Technical specifications of Macintosh Classic II from Apple's knowledge base (http://support.apple.com/kb/SP204) and from EveryMac.com (http://www.everymac.com/systems/apple/mac_classic/stats/mac_classic_ii.html). Retrieved June 24, 2010.

[54] Technical specifications of Macintosh LC II from Apple's knowledge base (http://support.apple.com/kb/SP206) and from EveryMac.com (http://www.everymac.com/systems/apple/mac_lc/stats/mac_lc_ii.html). Retrieved June 24, 2010.

[55] Technical specifications of Macintosh Quadra 700 from Apple's knowledge base (http://support.apple.com/kb/SP235) and from EveryMac.com (http://www.everymac.com/systems/apple/mac_quadra/stats/mac_quadra_700.html). Retrieved June 24, 2010.

[56] Technical specifications of Macintosh Quadra 900 from Apple's knowledge base (http://support.apple.com/kb/mac_quadra/stats/mac_quadra_900.html) and from EveryMac.com (http://www.everymac.com/systems/apple/mac_quadra/stats/mac_quadra_900.html). Retrieved June 24, 2010.

[57] Hornby, Thomas (January 3, 2005). "Apple's Transition to PowerPC put in perspective" (http://www.kaomso.com/FullStory.php?TheStory=78). Kaomso. . Retrieved December 24, 2007.

[58] Polsson, Ken (December 16, 2007). "Chronology of Apple Computer Personal Computers" (http://www.islandnet.com/~kpolsson/applehis/appl1994.htm). . Retrieved December 24, 2007.

[59] Polsson, Ken. "Chronology of Apple Computer Personal Computers" (http://www.islandnet.com/~kpolsson/applehis/appl1990.htm). . Retrieved November 18, 2007.

[60] Jade, Kasper (February 16, 2007). "Apple to re-enter the sub-notebook market" (http://www.appleinsider.com/articles/07/02/16/apple_to_re_enter_the_sub_notebook_market.html). AppleInsider. . Retrieved December 24, 2007.

[61] Technical specifications of PowerBook 165c from Apple's knowledge base (http://support.apple.com/kb/SP148) and from EveryMac.com (http://www.everymac.com/systems/apple/powerbook/stats/mac_powerbook165c.html). Retrieved June 24, 2010.

[62] Technical specifications of PowerBook 520 from Apple's knowledge base (http://support.apple.com/kb/SP160) and from EveryMac.com (http://www.everymac.com/systems/apple/powerbook/stats/mac_powerbook520.html). Retrieved November 12, 2010.

[63] Kunkel, Paul (October 1, 1997). *AppleDesign: The work of the Apple Industrial Design Group*. Rick English (photographs). New York City: Graphis Inc.. ISBN 1888001259.

[64] Apple Computer (June 19, 1995). "Macintosh Centris, Quadra 660AV: Description (Discontinued)" (http://docs.info.apple.com/article.html?artnum=12707). . Retrieved December 24, 2007.

[65] EveryMac.com (October 27, 2009). "Umax Mac Clones (MacOS-Compatible Systems)" (http://www.everymac.com/systems/umax/index-umax-supermac-mac-clones.html). . Retrieved November 11, 2009.

[66] EveryMac.com (October 27, 2009). "PowerComputing Mac Clones (MacOS-Compatible Systems)" (http://www.everymac.com/systems/powercc/index-power-computing-mac-clones.html). . Retrieved November 11, 2009.

[67] Knight, Dan (August 30, 2007). "1997: Apple Squeezes Mac Clones Out of the Market" (http://lowendmac.com/musings/mm07/0830.html). Low End Mac. . Retrieved December 24, 2007.

[68] Engst, Adam (January 23, 2009). "The six worst Apple products of all time" (http://www.macworld.com/article/138404/2009/01/macat25_worstproducts.html). Macworld. . Retrieved May 14, 2010.

[69] Technical specifications of iMac G3 from Apple's knowledge base (http://support.apple.com/kb/SP136) and from EveryMac.com (http://www.everymac.com/systems/apple/imac/stats/imac_ab.html). Retrieved June 24, 2010.

[70] "800,000 iMacs Sold in First 139 Days" (http://www.apple.com/ca/press/1999/01/iMac_Sales.html). Apple. January 5, 1999. . Retrieved December 23, 2007.

[71] Markoff, John (October 15, 1998). "COMPANY REPORTS; Apple's First Annual Profit Since 1995" (http://query.nytimes.com/gst/fullpage.html?res=9D0CE6D8123AF936A25753C1A96E958260). *The New York Times*. . Retrieved December 23, 2007.

[72] "iBook: An iMac to Go" (http://db.tidbits.com/article/5487?print_version=1). .

[73] "Apple Averages Three Thousand iBooks Per Day In Pre-orders!" (http://www.macobserver.com/news/99/august/990831/140000ibooks.html). The Mac Observer. August 31, 1999. . Retrieved December 24, 2007.

[74] "PC Data Ranks iBook Number One Portable in U.S." (http://www.apple.com/pr/library/2000/jan/25ibook.html). Apple. January 25, 2000. . Retrieved December 18, 2007.

[75] "Speed, Song Highlight Apple Product Announcements" (http://pcworld.about.com/news/Jan092001id37951.htm). .

[76] "Apple picks up the beat with CD-RW drives" (http://news.cnet.com/Apple-picks-up-the-beat-with-CD-RW-drives/2100-1040_3-249578.html). .

[77] <http://www.macworld.com/article/3744/2002/03/eisner2.html>

[78] "The Complete iTunes History – SoundJam MP to iTunes 9" (http://www.maclife.com/article/feature/complete_itunes_history_soundjam_mp_itunes_9). .

[79] "About the Macintosh Cube" (http://manuals.info.apple.com/en/PowerMacG4_CubeAbout.PDF) (PDF). Apple. 2000. . Retrieved October 9, 2008.

[80] Markoff, John; Hansell, Saul (January 12, 2005). "Apple Changes Course With Low-Priced Mac" (http://www.nytimes.com/2005/01/12/technology/12apple.html). *The New York Times*. . Retrieved January 16, 2006.

[81] *Apple unveils low-cost 'Mac mini'* (http://news.bbc.co.uk/1/hi/technology/4162009.stm). BBC News. January 11, 2005. . Retrieved April 28, 2010

[82] "Apple Developer Connection – Overview of the PowerPC System Software" (http://developer.apple.com/documentation/mac/PPCSoftware/PPCSoftware-12.html#HEADING12-0). Apple. . Retrieved May 11, 2009.

[83] Biersdorfer, J.D. (September 14, 2000). "Apple Breaks The Mold" (http://query.nytimes.com/gst/fullpage.html?res=9F01E2DD1538F937A2575AC0A9669C8B63). *The New York Times*. . Retrieved January 16, 2008.

[84] Technical specifications of MacBook Pro from Apple's knowledge base (http://support.apple.com/kb/SP36) and from EveryMac.com (http://www.everymac.com/systems/apple/macbook_pro/stats/macbook_pro_1.83.html). Retrieved June 24, 2010.

[85] "Apple to Use Intel Microprocessors Beginning in 2006" (http://www.apple.com/pr/library/2005/jun/06intel.html). Apple. June 6, 2005. . Retrieved May 14, 2010.

[86] Michaels, Philip (January 2, 2010). "Apple's most significant products of the decade" (http://www.macworld.com/article/145380/2010/01/10significantapplemoves.html). Macworld. . Retrieved May 14, 2010.

[87] "WWDC 2005 Keynote Live Update" (http://www.macworld.com/article/45157/2005/06/liveupdate.html). Macworld. June 6, 2005. . Retrieved May 14, 2010.

[88] "Hackers get Windows XP to run on a Mac" (http://www.msnbc.msn.com/id/11885495/). MSNBC (AP). March 17, 2006. . Retrieved April 24, 2006.
[89] "Boot Camp" (http://www.apple.com/support/bootcamp/). Apple. . Retrieved May 17, 2010.
[90] Dalrymple, Jim (March 5, 2006). "New Apple software lets Intel Macs boot Windows" (http://www.macworld.com/article/50204/2006/04/bootcamp.html). Macworld. . Retrieved May 14, 2010.
[91] "The story behind Apple's environmental footprint." (http://www.apple.com/environment/). Apple. . Retrieved January 24, 2011.
[92] "New MacBook Family Redefines Notebook Design" (http://www.apple.com/pr/library/2008/10/14macbook.html). Apple. October 14, 2008. . Retrieved December 23, 2009.
[93] Lam, Brian (October 20, 2009). "Apple iMac Hands On" (http://gizmodo.com/5385841/apple-imac-hands-on). *Gizmodo*. . Retrieved August 18, 2010.
[94] Kahney, Leander (June 25, 2003). "Design According to Ive" (http://www.wired.com/culture/design/news/2003/06/59381?currentPage=all). Wired. . Retrieved December 23, 2009.
[95] Nosowitz, Dan (November 7, 2009). "Watch Jonathan Ive's Segment in *Objectified*" (http://gizmodo.com/5399420/watch-jonathan-ives-segment-in-objectified). Gizmodo. . Retrieved December 23, 2009.
[96] "Apple Updates MacBook Pro Family with New Models & Innovative Built-in Battery for Up to 40% Longer Battery Life" (http://www.apple.com/pr/library/2009/06/08mbp.html). Apple. June 8, 2009. . Retrieved December 23, 2009.
[97] Apple Aims for the Masses With a Cheaper iPhone (http://www.nytimes.com/2008/06/10/technology/10apple.html)
[98] Windows or Mac? Apple Says Both (http://query.nytimes.com/gst/fullpage.html?res=9506EED81030F935A35757C0A9609C8B63&pagewanted=all)
[99] "Apple Reports First Quarter Results" (http://www.apple.com/pr/library/2010/01/25results.html). Apple. January 25, 2009. .
[100] "Apple's Mac market share grows 18.9% in first quarter" (http://www.loopinsight.com/2011/04/13/apples-mac-market-share-grows-18-9-in-first-quarter/). Loopinsight.com. April 13, 2011. . Retrieved July 5, 2011.
[101] "Thunderbolt Technology: The Fastest Data Connection to Your PC Just Arrived" (http://newsroom.intel.com/community/intel_newsroom/blog/2011/02/24/thunderbolt-technology-the-fastest-data-connection-to-your-pc-just-arrived). .
[102] Dvorak, John C. (March 18, 2003). "Apple Switch" (http://www.pcmag.com/article2/0,2817,933453,00.asp). *PC Magazine*. .
[103] "Apple Introduces the New iMac G5" (http://www.apple.com/pr/library/2005/oct/12imac.html). Apple. October 12, 2005. . Retrieved July 12, 2006.
[104] "Eek! A Two-Button Mac Mouse?" (http://www.wired.com/science/discoveries/news/2000/10/39773). Wired. October 30, 2000. . Retrieved December 23, 2009.
[105] "Apple Introduces Mighty Mouse" (http://www.apple.com/pr/library/2005/aug/02mightymouse.html). Apple. August 2, 2005. . Retrieved July 12, 2006.
[106] "Apple Debuts Wireless Mighty Mouse" (http://www.apple.com/pr/library/2006/jul/25mightymouse.html). Apple. July 25, 2006. . Retrieved November 30, 2007.
[107] "Apple Introduces Magic Mouse—The World's First Multi-Touch Mouse" (http://www.apple.com/pr/library/2009/10/20magicmouse.html). Apple. October 20, 2009. . Retrieved December 24, 2009.
[108] Welch, John (January 6, 2007). "Review: Mac OS X Shines In Comparison With Windows Vista" (http://www.informationweek.com/news/showArticle.jhtml?articleID=196800670&pgno=4). Information Week. . Retrieved February 5, 2007.
[109] Granneman, Scott (October 6, 2003). "Linux vs. Windows Viruses" (http://www.theregister.co.uk/2003/10/06/linux_vs_windows_viruses/). The Register. . Retrieved February 5, 2007.
[110] Gruber, John (June 4, 2004). "Broken Windows" (http://daringfireball.net/2004/06/broken_windows). Daring Fireball. . Retrieved April 24, 2006.
[111] "Operating System Market Share" (http://marketshare.hitslink.com/operating-system-market-share.aspx?qprid=8). September 2009. . Retrieved April 10, 2009.
[112] Roberts, Paul (February 21, 2006). "New Safari Flaw, Worms Turn Spotlight on Apple Security" (http://www.eweek.com/article2/0,1895,1929342,00.asp). eWeek. . Retrieved November 23, 2007.
[113] Conneally, Tim (August 28, 2009). "'Macs don't get viruses' myth dissolves before public's eyes" (http://www.betanews.com/article/Macs-dont-get-viruses-myth-dissolves-before-publics-eyes/1251493625). BetaNews.com. .
[114] "Apple security updates" (http://support.apple.com/kb/HT1222). Apple. January 21, 2009. . Retrieved January 29, 2009.
[115] Grimes, Roger A. (May 23, 2011). "7 questions about the Mac malware scare | Security" (http://www.infoworld.com/d/security/7-questions-about-the-mac-malware-scare-811). InfoWorld. . Retrieved July 5, 2011.
[116] "Mac Security Boasts Threatened by Malware Surge – International Business Times" (http://www.ibtimes.com/articles/152612/20110526/mac-security-boasts-threatened-by-malware-surge.htm). Ibtimes.com. May 26, 2011. . Retrieved July 5, 2011.
[117] Trenholm, Rich (May 20, 2011). "Apple tells support staff not to confirm Mac Defender infections" (http://crave.cnet.co.uk/software/apple-tells-support-staff-not-to-confirm-mac-defender-infections-50003832/). UK: Crave.. . Retrieved July 5, 2011.
[118] Seltzer, Larry (May 25, 2011). "Mac Defender 2.0 Released – Security Watch" (http://blogs.pcmag.com/securitywatch/2011/05/mac_defender_20_released.php). *PC Mag*. . Retrieved July 5, 2011.
[119] Lucas, Paul (June 4, 2005). "Paul J. Lucas's Mac Mini running Linux" (http://homepage.mac.com/pauljlucas/personal/macmini/index.html). . Retrieved December 23, 2009.

[120] Hoover, Lisa (April 11, 2008). "Virtualization Makes Running Linux a Snap" (http://ostatic.com/blog/virtualization-makes-running-linux-a-snap). . Retrieved December 23, 2009.

[121] Pogue, David; Joseph Schorr (1993). *Macworld Macintosh SECRETS*. San Mateo: IDG Books Worldwide, Inc.. p. 251. ISBN 1-56884-025-X.

[122] "Get a Mac advertisements" (http://www.apple.com/uk/getamac/). Apple. . Retrieved January 29, 2007.

[123] "Get a Mac" (http://www.apple.com/jp/getamac/). Apple. . Retrieved February 3, 2007.

[124] "Apple Announces Its Last Year at Macworld" (http://www.apple.com/pr/library/2008/12/16macworld.html). Apple. December 16, 2008. . Retrieved March 30, 2009.

[125] Polsson, Ken (July 29, 2009). "Chronology of Apple Computer Personal Computers" (http://www.islandnet.com/~kpolsson/applehis/appl1984.htm). . Retrieved August 27, 2009. See April 7, 1984.

[126] "Trends in Mac market share" (http://arstechnica.com/journals/apple.ars/2007/04/05/trends-in-mac-market-share). Ars Technica. April 5, 2009. . Retrieved August 27, 2009.

[127] "Apple Developer News, No. 87" (http://developer.apple.com/adcnews/pastissues/devnews121997.html#stats). Apple Computer. December 19, 1997. . Retrieved April 24, 2006.

[128] "Nearly 600 Million Computers-in-Use in Year 2000" (http://www.c-i-a.com/pr1198.htm). Computer Industry Almanac Inc. November 3, 1998. . Retrieved June 1, 2006.

[129] Dalrymple, Jim (April 20, 2005). "Apple desktop market share on the rise; will the Mac mini, iPod help?" (http://www.macworld.com/news/2005/03/20/marketshare/index.php). Macworld. . Retrieved April 24, 2006.

[130] Dalrymple, Jim (October 19, 2006). "Apple's Mac market share tops 5% with over 30% growth" (http://www.macworld.com/news/2006/10/19/marketshare/index.php). Macworld. . Retrieved December 22, 2006.

[131] "Operating System Market Share" (http://marketshare.hitslink.com/operating-system-market-share.aspx?qprid=8). Hitslink. July 2009. . Retrieved August 27, 2009.

[132] MacDailyNews (June 15, 2005). "16% of computer users are unaffected by viruses, malware because they use Apple Macs" (http://macdailynews.com/index.php/weblog/comments/5933/). . Retrieved April 24, 2006.

[133] (http://9to5mac.com/2011/03/17/top-10-mac-countries-by-market-share-united-states-is-3/)

[134] Gruber, John (July 23, 2003). "Market Share" (http://daringfireball.net/2003/07/market_share). Daring Fireball. . Retrieved April 24, 2006.

[135] Brockmeier, Joe (May 13, 2003). "What Will It Take To Put Apple Back on Top?" (http://www.newsfactor.com/perl/story/21499.html). NewsFactor Magazine online. . Retrieved April 24, 2006.

[136] "Despite growing sales, Mac's share of Apple profits wanes" (http://www.thenewstribune.com/2010/10/21/1390638/despite-growing-sales-macs-share.html). .

[137] Toporek, Chuck (August 22, 2001). "Apple, Market Share, and Who Cares?" (http://www.oreillynet.com/mac/blog/2001/08/apple_market_share_and_who_car.html). O'Reilly macdevcenter.com. . Retrieved April 24, 2006.

[138] Spero, Ricky (July 14, 2004). "Apple Posts Profit of $61 million; Revenue Jumps 30%" (http://www.macobserver.com/stockwatch/2004/07/14.1.shtml). The Mac Observer. . Retrieved April 24, 2006.

[139] Wilcox, Joe. "Macs Defy Windows' Gravity" (http://blogs.eweek.com/applewatch/content/channel/macs_defy_windows-gravity.html). Apple Watch. . Retrieved May 19, 2008.

[140] Fried, Ian (July 12, 2002). "Are Mac users smarter?" (http://news.com.com/2100-1040-943519.html). news.com. . Retrieved April 24, 2006.

## References

- Apple & Raskin, Jef (1992). *Macintosh Human Interface Guidelines*. Addison-Wesley Professional. ISBN 0-201-62216-5.
- Apple. "Press release Library" (http://www.apple.com/pr/library/). Retrieved November 18, 2007.
- Deutschman, Alan (2001). *The Second Coming of Steve Jobs*. Broadway. ISBN 0-7679-0433-8.
- Hertzfeld, Andy. "folklore.org: Macintosh stories" (http://folklore.org/index.py). Retrieved April 24, 2006.
- Hertzfeld, Andy (2004). *Revolution in the Valley*. O'Reilly Books. ISBN 0-596-00719-1.
- Kahney, Leander (2004). *The Cult of Mac*. No Starch Press. ISBN 1-886411-83-2.
- Kawasaki, Guy (1989). *The Macintosh Way*. Scott Foresman Trade. ISBN 0-673-46175-0.
- Kelby, Scott (2002). *Macintosh... The Naked Truth*. New Riders Press. ISBN 0-7357-1284-0.
- Knight, Dan (2005). "Macintosh History: 1984" (http://lowendmac.com/history/1984dk.shtml). Retrieved April 24, 2006.
- Levy, Steven (2000). *Insanely Great: The Life and Times of Macintosh, the Computer That Changed Everything*. Penguin Books. ISBN 0-14-029177-6.
- Linzmayer, Owen (2004). *Apple Confidential 2.0*. No Starch Press. ISBN 1-59327-010-0.

- Page, Ian (2007). "MacTracker Macintosh model database 4.3.1" (http://www.mactracker.ca/). Retrieved November 31, 2007.
- Sanford, Glen (2006). "Apple History" (http://www.apple-history.com/). Retrieved April 24, 2006.
- Singh, Amit (2005). "A History of Apple's Operating Systems" (http://www.kernelthread.com/mac/oshistory/). Retrieved April 24, 2006.


## External links

- Official website (http://http://www.apple.com/mac/)
- Making the Macintosh: Technology and Culture in Silicon Valley (http://library.stanford.edu/mac/index.html)

# Audi

| | |
|---|---|
| **Type** | Private company<br>(FWB Xetra: NSU [1]) |
| **Industry** | Automotive industry |
| **Founded** | Zwickau, Germany (16 July 1909)[2] |
| **Founder(s)** | August Horch |
| **Headquarters** | Ingolstadt, Germany |
| **Number of locations** | *Production locations:*<br>Germany:<br>Ingolstadt & Neckarsulm<br>Hungary: Győr<br>Belgium: Brussels<br>China: Changchun<br>India: Aurangabad<br>Brazil: Curitiba |
| **Area served** | Worldwide |
| **Key people** | Rupert Stadler<br>Chairman of the Board of Management,<br>Wolfgang Egger<br>Head of Design |
| **Products** | Automobiles, Engines |
| **Production output** | ▲ 1,143,902 units (*2010*)<br>(only Audi brand) |
| **Revenue** | €35.441 billion (*2010*)<br>(US$52.57 billion USD)<br>(including subsidiaries) |
| **Profit** | €1.850 billion (*2009*)<br>(US$2.74 billion USD) |
| **Total assets** | €16.832 billion (*2009*)<br>(US$25 billion USD) |
| **Total equity** | €3.451 billion (*2009*)<br>(US$5.12 billion USD) |
| **Employees** | 46,372 (2009)[3] |
| **Parent** | Volkswagen Group |

| Subsidiaries | • Audi do Brasil e Cia (Curitiba, Brazil)<br>• Audi Hungaria Motor Kft. (Győr, Hungary)<br>• Audi Senna Ltda. (Brazil)<br>• Automobili Lamborghini Holding S.p.A (Sant'Agata Bolognese, Italy)<br>• Autogerma S.p.A. (Verona, Italy)<br>• Ducati Motor Holding S.p.A. (Bologna, Italy)<br>• quattro GmbH (Neckarsulm, Germany)<br>• Audi India (Aurangabad, India) |
|---|---|
| Website | audi.com [4] |

**Audi AG** (Xetra: NSU [1]) is a German automobile manufacturer, from supermini to crossover SUVs in various body styles and price ranges that are marketed under the **Audi** brand (German pronunciation: [ˈaʊdi]), positioned as the premium brand within the Volkswagen Group.[5]

The company is headquartered in Ingolstadt, Bavaria, Germany, and has been a wholly owned (99.55%)[6] subsidiary of Volkswagen AG since 1966, following a phased purchase of its predecessor, Auto Union, from its former owner, Daimler-Benz. Volkswagen relaunched the Audi brand with the 1965 introduction of the Audi F103 series.

The company name is based on the surname of the founder August Horch, his surname meaning *listen* in German—which, when translated into Latin, becomes *Audi*. The four rings each represent individual car companies that banded together to create the union. Its slogan or tagline is Vorsprung durch Technik, meaning "Progress through Technology". More recently, however, Audi has updated the tagline to "Truth in Engineering" in the U.S.

## Within Volkswagen Group

The largest shareholder of Audi AG is Volkswagen AG, which holds over 99 percent of the share capital. Volkswagen AG includes the consolidated accounts of Audi AG in its own financial statements. In recent years, the possibility of Audi being spun off or otherwise divested by Volkswagen has been mooted.[7]

From 2002 up to 2007, Audi headed the *Audi Brand Group*, the Volkswagen Group's automotive sub-division, consisted of Audi together with SEAT and Lamborghini, that was focused on more sporty values, with the marques' product vehicles and performance being under the higher responsibility of the Audi brand.

Audi Australia are said to be lacking in ability to help customers who are being bullied by Audi service centre's, according to Audi Australia, they have nothing to do with any complaints or anything in that matter. They tend to leave their customers in the lurch. The focus on money is enourmous and the main aim for many Audi Australia franchises is to seep as much money as possible out of your back pocket. Many thousands of complaints are made againts Audi Australia franchises and many angry customers are leaving Audi in the lurch and choosing more reliable and nice service.

Customers report the feeling of being spoken down to, especially those who own the older models, as Audi centres tend to try and make sure it costs you in the hope of you taking a loan out to buy a brand new Audi.

With the exception of some Audi Centres, the majority of Audi Centres are the same, Areas such as Newcastle are said to be the ones to steer clear of. [8] [9] [10]

# History

## Birth of the company and its name

On 14 November 1899, August Horch (1868–1951) established the company A. Horch & Cie. in the Ehrenfeld district of Cologne, but because of a dispute between him and the supervisory board, he had to leave the company. In 1909, he established a second company, Horch Automobil-Werke GmbH. The first Audi automobile, the Audi Type A 10/22 hp (**unknown operator: u'strong'** kW) Sport-Phaeton, was produced in 1910 in Zwickau.[11]

In 1909, Horch was forced out of the company he had founded.[12] He then started a new company in Zwickau and continued using the Horch brand. His former partners sued him for trademark infringement, and the German Supreme Court (Reichsgericht in Berlin),[13] finally determined that the Horch brand belonged to his former company.[12]

Audi Type E

Since August Horch was barred from using "Horch" as a trade name in his new car business, he called a meeting with close business friends, Paul and Franz Fikentscher from Zwickau, Germany. At the apartment of Franz Fikentscher, they discussed how to come up with a new name for the company. During this meeting, Franz's son was quietly studying Latin in a corner of the room. Several times he looked like he was on the verge of saying something but would just swallow his words and continue working, until he finally blurted out, "Father – *audiatur et altera pars...* wouldn't it be a good idea to call it *audi* instead of *horch*?"[14] "Horch!" in German means "Hark!" or "hear", which is "Audi" in the singular imperative form of "audire" - "to listen" - in Latin. The idea was enthusiastically accepted by everyone attending the meeting.[15] The first Audi car, the Type B, 10/28PS was delivered early in 1910.[16]

Audi started with a 2,612 cc inline-four engine model followed by a 3,564 cc model, as well as 4,680 cc and 5,720 cc models. These cars were successful even in sporting events. The first six-cylinder model, 4,655 cc appeared in 1924.

August Horch left the Audi company in 1920 for a high position at the ministry of transport, but he was still involved with Audi as a member of the board of trustees. In September 1921, Audi became the first German car manufacturer to present a production car, the Audi Type K, with left-handed drive.[17] Left-hand drive spread and established dominance during the 1920s because it provided a better view of oncoming traffic, making overtaking safer.[17]

## Auto Union era

In August 1928, Jørgen Rasmussen, the owner of Dampf-Kraft-Wagen (DKW), acquired the majority of shares in Audiwerke AG.[18] In the same year, Rasmussen bought the remains of the U.S. automobile manufacturer Rickenbacker, including the manufacturing equipment for eight-cylinder engines. These engines were used in *Audi Zwickau* and *Audi Dresden* models that were launched in 1929. At the same time, six-cylinder and four-cylinder (licensed from Peugeot) models were manufactured. Audi cars of that era were luxurious cars equipped with special bodywork.

In 1932, Audi merged with Horch, DKW, and Wanderer, to form Auto Union. It was during this period that the company offered the Audi Front that became the first European car to combine a six-cylinder engine with front-wheel drive. It used a powertrain shared with the Wanderer, but turned 180-degrees, so that the drive shaft faced the front.

Before World War II, Auto Union used the four interlinked rings that make up the Audi badge today, representing these four brands. This badge was used, however, only on Auto Union racing cars in that period while the member companies used their own names and emblems. The technological development became more and more concentrated and some Audi models were propelled by Horch or Wanderer built engines.

Reflecting the economic pressures of the time, Auto Union concentrated increasingly on smaller cars through the 1930s, so that by 1938 the company's DKW brand accounted for 17.9% of the German car market, while Audi held only 0.1%.

## Post-World War II

Like most German manufacturing, at the onset of World War II the Auto Union plants were retooled for military production, and were a target for allied bombing during the war, leaving them damaged.

Overrun by the Soviet Army in 1945, on the orders of the Soviet Union military administration the factories were dismantled as part of war reparations.[19] Following this, the company's entire assets were expropriated without compensation.[19] On 17 August 1948, Auto Union AG of Chemnitz was deleted from the commercial register.[18] These actions had the effect of liquidating Germany's Auto Union AG. The remains of the Audi plant of Zwickau became the VEB (for "People Owned Enterprise") Automobilwerk Zwickau or AWZ (in English: Automobile Works Zwickau).[20]

IFA F9

The former Audi factory in Zwickau restarted assembly of the pre-war-models in 1949. These DKW models were renamed to IFA F8 and IFA F9 and were similar to the West German versions. West and East German models were equipped with the traditional and renowned DKW two-stroke engines.

## New Auto Union unit

A new West German headquartered Auto Union was launched in Ingolstadt, Bavaria with loans from the Bavarian state government and Marshall Plan aid.[21] The reformed company was launched 3 September 1949 and continued DKW's tradition of producing front-wheel drive vehicles with two-stroke engines.[19] This included production of a small but sturdy 125 cc motorcycle and a DKW delivery van, the DKW F 89 L at Ingolstadt. The Ingolstadt site was large, consisting of an extensive complex of formerly military buildings which was suitable for administration as well as vehicle warehousing and distribution, but at this stage there was at Ingolstadt no dedicated plant suitable for mass production of automobiles: for manufacturing the company's first post-war mass-market passenger car plant capacity in Düsseldorf was rented from Rheinmetall-Borsig. It was only ten years later, after the company had attracted an investor that funds became available for construction of major car plant at the Ingolstadt head office site.

In 1958, Daimler-Benz took an 87% holding in the Auto Union company, and this was increased to a 100% holding in 1959. However, small two-stroke cars were not the focus of Daimler-Benz's interests, and while the early 1960s saw major investment in new Mercedes models and in a state of the art factory for Auto Union's, the company's ageing model range at this time did not benefit from the economic boom of the early 1960s to the same extent as competitor manufacturers such as Volkswagen and Opel. The decision to dispose of the Auto Union business was based on its lack of profitability.[22] Ironically, by the time they sold the business it also included a large new factory and near production-ready modern four-stroke engine, which would enable the Auto Union business, under a new owner and with the benefit of a rediscovered name, Audi. Under the terms of the sale, Daimler-Benz retained the old Düsseldorf plant, which survives to the present day as a centre for Mercedes-Benz commercial vehicle assembly.

In 1964, Volkswagen acquired a 50% holding in the business, which included the new factory in Ingolstadt and the trademark rights of the Auto Union. Eighteen months later, Volkswagen bought complete control of Ingolstadt, and by 1966 were using the spare capacity of the Ingolstadt plant to assemble an additional 60,000 Volkswagen Beetles per year.[23] Two-stroke engines became less popular during the 1960s as customers were more attracted to the smoother four-stroke engines. In September 1965, the DKW F102 was fitted with a four-stroke engine and a facelift for the car's front and rear. Volkswagen dumped the DKW brand because of its associations with two-stroke technology, and having classified the model internally as the F103, sold it simply as the "Audi." Later developments of the model were named after their horsepower ratings and sold as the Audi 60, 75, 80, and Super 90, selling until 1972. Initially, Volkswagen was hostile to the idea of Auto Union as a standalone entity producing its own models having acquired the company merely to boost its own production capacity through the Ingolstadt assembly plant. Then VW chief Heinz Nordhoff explicitly forbode Auto Union from any further product development. Fearing the the company's heritage would disappear underneath VW badge engineering, Auto Union engineers under the leadership of Ludwig Kraus developed the first Audi 100 in secret, without Nordhoff's knowledge. When presented with a finished prototype, Nordhoff was so impressed he authorised the car for production, which when launched in 1968, went on to be a huge success. With this, the resurrection of the Audi brand was now complete, this being followed by the first generation Audi 80 in 1972, which would in turn provide the impetus for VW's new front wheel drive watercooled range which debuted from the mid 1970s onward.

Audi 80 assembly line in Wolfsburg, 1973

In 1969, Auto Union merged with NSU, based in Neckarsulm, near Stuttgart. In the 1950s, NSU had been the world's largest manufacturer of motorcycles, but had moved on to produce small cars like the NSU Prinz, the TT and TTS versions of which are still popular as vintage race cars. NSU then focused on new rotary engines based on the ideas of Felix Wankel. In 1967, the new NSU Ro 80 was a car well ahead of its time in technical details such as aerodynamics, light weight, and safety. However, teething problems with the rotary engines put an end to the independence of NSU. The Neckarsulm plant is now used to produce the larger Audi models A6 and A8. The Neckarsulm factory is also home of the quattro GmbH, a subsidiary responsible for development and production of Audi high-performance models: the R8 and the **"RS"** model range.

The mid-sized car that NSU had been working on, the K70, was intended to slot between the rear-engined Prinz models and the futuristic NSU Ro 80. However, Volkswagen took the K70 for its own range, spelling the end of NSU as a separate brand.

## Modern era

The new merged company was known as **Audi NSU Auto Union AG**, and saw the emergence of Audi as a separate brand for the first time since the pre-war era. Volkswagen introduced the Audi brand to the United States for the 1970 model year.

The first new car of this regime was the Audi 100 of 1968. This was soon joined by the Audi 80/Fox (which formed the basis for the 1973 Volkswagen Passat) in 1972 and the Audi 50 (later rebadged as the Volkswagen Polo) in 1974. The Audi 50 was a seminal design because it was the first incarnation of the Golf/Polo concept, one that led to a hugely successful world car.

The Audi image at this time was a conservative one, and so, a proposal from chassis engineer Jörg Bensinger[24] was accepted to develop the four-wheel drive technology in Volkswagen's Iltis military vehicle for an Audi performance car and rally racing car. The performance car, introduced in 1980, was named the "Audi Quattro", a turbocharged coupé which was also the first German large-scale production vehicle to feature permanent all-wheel drive through a centre differential. Commonly referred to as the "Ur-Quattro" (the "Ur-" prefix is a German augmentative used, in this case, to mean "original" and is also applied to the first generation of Audi's S4 and S6 Sport Saloons, as in "UrS4" and "UrS6"), few of these vehicles were produced (all hand-built by a single team), but the model was a great success in rallying. Prominent wins proved the viability of all-wheel drive racecars, and the Audi name became associated with advances in automotive technology.

In 1985, with the Auto Union and NSU brands effectively dead, the company's official name was now shortened to simply **Audi AG**.

In 1986, as the Passat-based Audi 80 was beginning to develop a kind of "grandfather's car" image, the *type 89* was introduced. This completely new development sold extremely well. However, its modern and dynamic exterior belied the low performance of its base engine, and its base package was quite spartan (even the passenger-side mirror was an option.) In 1987, Audi put forward a new and very elegant Audi 90, which had a much superior set of standard features. In the early 1990s, sales began to slump for the Audi 80 series, and some basic construction problems started to surface.

Audi Quattro

In the early part of the 21st century, Audi set forth on a German racetrack to claim and maintain several world records, such as top speed endurance. This effort was in-line with the company's heritage from the 1930s racing era Silver Arrows.

Through the early 1990s, Audi began to shift its target market upscale to compete against German automakers Mercedes-Benz and BMW. This began with the release of the Audi V8 in 1990. It was essentially a new engine fitted to the Audi 100/200, but with noticeable bodywork differences. Most obvious was the new grille that was now incorporated in the bonnet.

By 1991, Audi had the four-cylinder Audi 80, the 5-cylinder Audi 90 and Audi 100, the turbocharged Audi 200 and the Audi V8. There was also a coupe version of the 80/90 with both 4- and 5-cylinder engines.

Although the five-cylinder engine was a successful and robust powerplant, it was still a little too different for the target market. With the introduction of an all-new Audi 100 in 1992, Audi introduced a 2.8L V6 engine. This engine was also fitted to a face-lifted Audi 80 (all 80 and 90 models were now badged 80 except for the USA), giving this model a choice of four-, five-, and six-cylinder engines, in Saloon, Coupé and Cabriolet body styles.

The five-cylinder was soon dropped as a major engine choice; however, a turbocharged 230 hp (169 kW) version remained. The engine, initially fitted to the 200 quattro 20V of 1991, was a derivative of the engine fitted to the Sport Quattro. It was fitted to the Audi Coupé, and named the S2 and also to the Audi 100 body, and named the S4. These two models were the beginning of the mass-produced S series of performance cars.

## Audi 5000 unintended acceleration

Sales in the United States fell after a series of recalls from 1982 to 1987 of Audi 5000 models[25] associated with reported incidents of sudden unintended acceleration linked to six deaths and 700 accidents.[25] At the time, NHTSA was investigating 50 car models from 20 manufacturers for sudden surges of power.[26]

A *60 Minutes* report aired 23 November 1986,[27] featuring interviews with six people who had sued Audi after reporting unintended acceleration, showing an Audi 5000 ostensibly suffering a problem when the brake pedal was pushed.[28] [29] Subsequent investigation revealed that *60 Minutes* had engineered the failure – fitting a canister of compressed air on the passenger-side floor, linked via a hose to a hole drilled into the transmission.[27]

Audi contended, prior to findings by outside investigators,[26] that the problems were caused by driver error, specifically pedal misapplication.[26] Subsequently, the National Highway Traffic Safety Administration (NHTSA) concluded that the majority of unintended acceleration cases, including all the ones that prompted the *60 Minutes* report, were caused by driver error such as confusion of pedals.[30] CBS did not acknowledge the test results of involved government agencies, but did acknowledge the similar results of another study.[28]

Audi 100 C3, sold as the Audi 5000 in the U.S.

With the series of recall campaigns, Audi made several modifications; the first adjusted the distance between the brake and accelerator pedal on automatic-transmission models.[25] Later repairs, of 250,000 cars dating back to 1978, added a device requiring the driver to press the brake pedal before shifting out of park.[25] A legacy of the Audi 5000 and other reported cases of sudden unintended acceleration are intricate gear stick patterns and brake interlock mechanisms to prevent inadvertent shifting into forward or reverse.

Audi's U.S. sales, which had reached 74,061 in 1985, dropped to 12,283 in 1991 and remained level for three years.[25] – with resale values falling dramatically.[31] Audi subsequently offered increased warranty protection[31] and renamed the affected models – with the *5000* becoming the *100* and *200* in 1989[26] – and only reached the same sales levels again by model year 2000.[25]

A 2010 *BusinessWeek* article – outlining possible parallels between Audi's experience and 2009–2010 Toyota vehicle recalls – noted a class-action lawsuit filed in 1987 by about 7,500 Audi 5000-model owners remains unsettled and is currently being contested in county court in Chicago after appeals at the Illinois state and U.S. federal levels.[25]

## Model introductions

In the mid-to-late 1990s, Audi introduced new technologies including the use of aluminum construction. Produced from 1999 to 2005, the Audi A2 was a futuristic super mini, born from the Al2 concept, with many features that helped regain consumer confidence, like the aluminium space frame, which was a first in production car design. In the A2 Audi further expanded their TDI technology through the use of frugal three-cylinder engines. The A2 was extremely aerodynamic and was designed around a wind tunnel. The Audi A2 was criticised for its high price and was never really a sales success but it planted Audi as a cutting-edge manufacturer. The model, a Mercedes-Benz A-Class competitor, sold relatively well in Europe. However, the A2 was discontinued in 2005 and Audi decided not to develop an immediate replacement.

The next major model change came in 1995 when the Audi A4 replaced the Audi 80. The new nomenclature scheme was applied to the Audi 100 to become the Audi A6 (with a minor facelift). This also meant the S4 became the S6 and a new S4 was introduced in the A4 body. The S2 was discontinued. The Audi Cabriolet continued on (based on the Audi 80 platform) until 1999, gaining the engine upgrades along the way. A new A3 hatchback model (sharing the Volkswagen Golf Mk4's platform) was introduced to the range in 1996, and the radical Audi TT coupé and roadster were debuted in 1998 based on the same underpinnings.

The engines available throughout the range were now a 1.4 L, 1.6 L and 1.8 L four-cylinder, 1.8 L four-cylinder turbo, 2.6 L and 2.8 L V6, 2.2 L turbo-charged five-cylinder and the 4.2 L V8 engine. The V6s were replaced by new 2.4 L and 2.8 L 30V V6s in 1998, with marked improvement in power, torque and smoothness. Further engines were added along the way, including a 3.7 L V8 and 6.0 L W12 engine for the A8.

## Audi AG today

Audi plant in Aurangabad, India

Audi's sales grew strongly in the 2000s, with deliveries to customers increasing from 653,000 in 2000 to 1,003,000 in 2008. The largest sales increases came from Eastern Europe (+19.3%), Africa (+17.2%) and the Middle East (+58.5%). China in particular has become a key market, representing 108,000 out of 705,000 cars delivered in the first three quarters of 2009. One factor for its popularity in China is that Audis have become the car of choice for purchase by the Chinese government for officials, and purchases by the government are responsible for 20% of its sales in China.[32] As of late 2009, Audi's operating profit of €1.17-billion ($1.85-billion) made it the biggest contributor to parent Volkswagen Group's nine-month operating profit of €1.5-billion, while the other marques in Group such as Bentley and Seat had suffered considerable losses.[33]

Audi has 7 manufacturing plants around the world[34] , although many sub-assemblies such as engines and transmissions are manufactured within other Volkswagen Group plants around the world.

Audi's two principal assembly plants are:

- Ingolstadt, Opened by Auto Union in 1964, (A3, A4, A5, Q5)
- Neckarsulm, Acquired from NSU in 1969 (A6, A7, A8, R8 & all RS variants)

Outside of Germany, Audi produces vehicles at:

- Bratislava, shared with Volkswagen and Porsche (Q7)
- Martorell, shared with SEAT and Volkswagen (Q3)
- Györ, Hungary, (TT and some A3 variants)
- Brussels, Belgium (A3 Cabriolet, A1)
- Changchun, China since 1995,
- Aurangabad, India since 2006.

# Technology

## Vehicle bodies

### Bodyshells

Audi produces 100% galvanised cars to prevent corrosion,[35] and was the first mass-market vehicle to do so, following introduction of the process by Porsche, c.1975. Along with other precautionary measures, the full-body zinc coating has proved to be very effective in preventing rust. The body's resulting durability even surpassed Audi's own expectations, causing the manufacturer to extend its original 10-year warranty against corrosion perforation to currently 12 years (except for aluminium bodies which do not rust).[36]

### Space frame

Audi introduced a new series of vehicles in the mid-1990s and continues to new technology and high performance. An all-aluminium car was brought forward by Audi, and in 1994 the Audi A8 was launched, which introduced aluminium space frame technology (called *Audi Space Frame* or ASF) which saves weight and improves torsion rigidity compared to a conventional steel frame. Prior to that effort, Audi used examples of the Type 44 chassis fabricated out of aluminium as test-beds for the technique. The disadvantage of the aluminium frame is that it is very expensive to repair and requires a specialized aluminium bodyshop.[37] The weight reduction is somewhat offset by the quattro four-wheel drive system which is standard in most markets. Nonetheless, the A8 is usually the lightest all-wheel drive car in the full-size luxury segment, also having best-in-class fuel economy.[38] The Audi A2, Audi TT and Audi R8 also use Audi Space Frame designs.

## Drivetrains

### Layout

Audi R8

In all its post Volkswagen-era models, Audi has firmly refused to adopt the traditional rear-wheel drive layout favored by its two arch rivals Mercedes-Benz and BMW, favoring either front-wheel drive or all-wheel drive. The majority of Audi's lineup in the United States features all-wheel drive standard on most of its expensive vehicles (only the entry-level trims of the A4 and A6 are available with front-wheel drive), in contrast to Mercedes-Benz and BMW whose lineup treats all-wheel drive as an option. BMW only offered all-wheel drive on its V8-powered cars (as opposed to crossover SUVs)(Needs to be checked - BMW X-Drive is available on 6 cylinder models - see BMW X-Drive (http:/ / en. wikipedia. org/ wiki/ BMW_xDrive) starting with the 2010 BMW 7 Series and 2011 BMW 5 Series, while the Audi A8 has had all-wheel drive available/standard since the 1990s. Regarding high-performance variants, Audi S and RS models have always had all-wheel drive, unlike their direct rivals from BMW M and Mercedes-AMG whose cars are rear-wheel drive only (although their performance crossover SUVs are all-wheel drive).

For most of its lineup (excluding the A3, A1, and TT models), Audi has not adopted the transverse engine layout which is typically found in economy cars (as well as more upscale European manufacturers such as Peugeot and Citroën), since that would limit the type and power of engines that can be installed. In order to be able to mount powerful engines (such as a V8 engine in the Audi S4 and Audi RS4), Audi has usually engineered its more expensive cars with a longitudinally front-mounted engine, in an "overhung" position, over the front wheels in front of the axle line. While this allows for the easy adoption of all-wheel drive, it goes against the ideal 50:50 weight distribution (as do all front wheel drive cars).

Audi has recently applied the *quattro* badge to models such as the A3 and TT which do not use the Torsen-based system as in prior years with a mechanical center differential, but with the Swedish Haldex Traction electro-mechanical clutch AWD system.

### Engines

Further information: Audi production engines

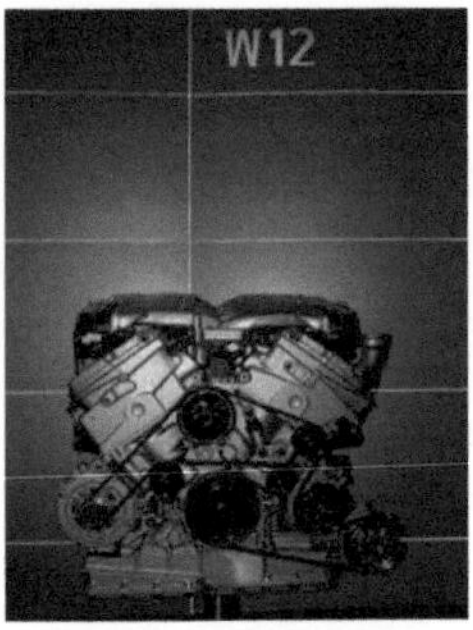

Volkswagen Group W12 engine from the Volkswagen Phaeton W12

In the 1980s, Audi, along with Volvo, was the champion of the inline-five cylinder, 2.1/2.2 L engine as a longer-lasting alternative to more traditional six-cylinder engines. This engine was used not only in production cars but also in their race cars. The 2.1 L inline five-cylinder engine was used as a base for the rally cars in the 1980s, providing well over 400 horsepower (298 kW) after modification. Before 1990, there were engines produced with a displacement between 2.0 L and 2.3 L. This range of engine capacity allowed for both fuel economy and power.

For the ultra-luxury version of its Audi A8 fullsize luxury flagship sedan, the Audi A8L W12, Audi uses the Volkswagen Group W12 engine instead of the conventional V12 engine favored by rivals Mercedes-Benz and BMW. The W12 engine configuration (also known as a "WR12") is created by forming two imaginary narrow-angle 15° VR6 engines at an angle of 72°, and the narrow angle of each set of cylinders allows just two overhead camshafts to drive each pair of banks, so just four are needed in total.[39] The advantage of the W12 engine is its compact packaging, allowing Audi to build a 12-cylinder sedan with all-wheel drive, whereas a conventional V12 engine could only have a rear-wheel drive configuration as it would have no space in the engine bay for a differential and other components required to power the front wheels. In fact, the 6.0 L W12 in the Audi A8L W12 is smaller in overall dimensions than the 4.2 L V8 that powers the Audi A8 4.2 variants. [40] The 2011 Audi A8 debuted a revised 6.3-litre version of the W12 (WR12) engine with 500 PS (**unknown operator: u'strong'** kW; **unknown operator: u'strong'** hp).

### Fuel Stratified Injection

New models of the A3, A4, A6 and A8 have been introduced, with the ageing 1.8-litre engine now having been replaced by new Fuel Stratified Injection (FSI) engines. Nearly every petroleum burning model in the range now incorporates this fuel-saving technology.

### Direct-Shift Gearbox

V8 FSI engine.

At the turn of the century, Volkswagen introduced the Direct-Shift Gearbox (DSG), a type of dual clutch transmission. It is an automated semi-automatic transmission, drivable like a conventional automatic transmission. Based on the gearbox found in the Group B S1, the system includes dual electrohydraulically controlled clutches instead of a torque converter. This is implemented in some VW Golfs, Audi A3 and TT models where DSG is called S-tronic.

### LED daytime running lights

Beginning in 2006, Audi has implemented white LED technology as daytime running lights in their products. The distinctive shape of the DRLs has become a trademark of sorts. LEDs were first introduced on the Audi S6, and have since spread throughout the entire model range. The LEDs are present on some Audi billboards.

Since 2010, Audi has also offered the LED technology in low- and high-beam headlights.[41]

### Multi Media Interface

The DRL in a Audi A4 B8

Audi has recently started offering a computerised control system for its cars, called Multi Media Interface (MMI). This came amid criticism of BMW's iDrive control. It is essentially a rotating control knob and 'segment' buttons – designed to control all in-car entertainment devices (radio, CD changer, iPod, TV tuner), satellite navigation, heating and ventilation, and other car controls with a screen. MMI was widely reported to be a considerable improvement on BMW's iDrive, although BMW has since made their iDrive more user-friendly.

MMI has been generally well-received, as it requires less menu-surfing with its segment buttons around a central knob, along with 'main function' direct access buttons – with shortcuts to the radio or phone functions. The screen, either colour or monochrome, is mounted on the upright dashboard, and on the A4 (new), A5, A6, A8, and Q7, the controls are mounted horizontally.

An "MMI-like" system is also available on the A3, TT, A4 (B7), and R8 models – when equipped with the Audi Navigation System Plus (RNS-E) satellite navigation system.

## Models

### Current model range

The following tables list Audi production vehicles that are sold as of 2012:

**Audi cars**

| | | | |
|---|---|---|---|
| A1 |  | Supermini | • 3-door Hatchback<br>• Sportback (5-door Hatchback) |
| A3 |  | Small Family Car | • 3-door Hatchback<br>• Sportback (5-door Hatchback)<br>• Cabriolet (Convertible) |
| A4 |  | Compact Executive Car | • Saloon (Sedan)<br>• Avant (Estate/Wagon)<br>• Allroad (Crossover Estate/Wagon) |
| A5 |  | Compact Executive Car | • Coupé<br>• Sportback (5-door Hatchback)<br>• Cabriolet (Convertible) |
| A6 |  | Executive Car | • Saloon (Sedan)<br>• Avant (Estate/Wagon)<br>• Allroad (Crossover Estate/Wagon) |
| A7 |  | Executive Car | • Sportback (5-door Hatchback) |
| A8 |  | Full-size Luxury Car | • Saloon (Sedan) |

## Audi coupés and SUVs

| | | | |
|---|---|---|---|
| TT |  | Compact Sports Car | • Coupé<br>• Roadster (Convertible) |
| R8 |  | Sports Car | • Coupé<br>• Spyder (Convertible) |
| Q3 |  | Compact Crossover SUV | • SUV |
| Q5 |  | Compact Crossover SUV | • SUV |
| Q7 |  | Full-size Crossover SUV | • SUV |

## S and RS models

### S (Sport) models

| | | | |
|---|---|---|---|
| S3 |  | Small Family Car | • 3-door Hatchback<br>• Sportback (5-door Hatchback) |
| S4 |  | Compact Executive Car | • Saloon (Sedan)<br>• Avant (Estate/Wagon) |
| S5 |  | Compact Executive Car | • Coupé |
| S6[42] |  | Executive Car | • Saloon (Sedan)<br>• Avant (Estate/Wagon) |
| S7[43] | | Executive Car | • Sportback (5-door Hatchback) |
| S8[44] | | Full-size Luxury Car | • Saloon (Sedan) |
| TTS |  | Compact Sports Car | • Coupé<br>• Roadster (Convertible) |

### RS (RennSport/Racing) models

| | | | |
|---|---|---|---|
| RS3 |  | Small Family Car | • 3-door Hatchback<br>• Sportback (5-door Hatchback) |
| RS4 |  | Compact Executive Car | • Avant (Estate/Wagon) |
| RS5 |  | Compact Executive Car | • Coupé |
| TT RS |  | Compact Sports Car | • Coupé<br>• Roadster (Convertible) |

### Electric vehicles

Further information: Audi concept vehicles

Audi is planning an alliance with the Japanese electronic giant Sanyo to develop a pilot hybrid electric project for the Volkswagen Group. The alliance could result in Sanyo batteries and other electronic components being used in future models of the Volkswagen Group.[45] Concept electric vehicles unveiled to date include the Audi A1 Sportback Concept,[46] Audi A4 TDI Concept E,[47] and the fully electric Audi e-tron Concept Supercar.[48]

## Production figures

| | A1 | A2 | A3 | A4 | A5 | A6 | A7 | A8 | Q3 | Q5 | Q7 | TT | R8 |
|---|---|---|---|---|---|---|---|---|---|---|---|---|---|
| [49] 1998 | — | — | 143,974 | 271,152 | — | 174,867 | — | 15,355 | — | — | — | 13,682 | — |
| [49] 1999 | — | — | 143,505 | 252,514 | — | 162,573 | — | 14,636 | — | — | — | 52,579 | — |
| [50] 2000 | — | 32,164 | 136,141 | 231,869 | — | 180,715 | — | 12,894 | — | — | — | 56,776 | — |
| [51] 2001 | — | 49,369 | 131,082 | 308,778 | — | 186,467 | — | 11,708 | — | — | — | 39,349 | — |
| [52] 2002 | — | 37,578 | 125,538 | 360,267 | — | 178,773 | — | 10,942 | — | — | — | 34,711 | — |
| [53] 2003 | — | 27,323 | 159,417 | 353,836 | — | 168,612 | — | 21,748 | — | — | — | 32,337 | — |
| [54] 2004 | — | 19,745 | 181,274 | 345,231 | — | 195,529 | — | 22,429 | — | — | — | 23,605 | — |
| [55] 2005 | — | 10,026 | 224,961 | 337,705 | — | 215,437 | — | 21,515 | — | — | 1,185 | 12,307 | — |
| [56] 2006 | — | — | 231,752 | 341,110 | 487 | 229,021 | — | 22,468 | — | — | 72,169 | 23,675 | 164 |
| [57] 2007 | — | — | 231,117 | 289,806 | 25,549 | 243,842 | — | 22,182 | — | 162 | 77,395 | 56,766 | 4,125 |
| [58] 2008 | — | — | 222,164 | 378,885 | 57,650 | 214,074 | — | 20,140 | — | 20,324 | 59,008 | 41,789 | 5,656 |
| [59] 2009 | — | — | 206,747 | 282,033 | 84,883 | 182,090 | — | 8,599 | — | 105,074 | 27,929 | 22,821 | 2,101 |
| [60] 2010 | 51,937 | — | 198,974 | 306,291 | 111,270 | 211,256 | 8,496 | 22,435 | — | 154,604 | 48,937 | 26,217 | 3,485 |

- Data from 1998 to 2010. Figures for different body types/versions of models have been merged to create overall figures for each model.

## Motorsports

Audi has competed in numerous forms of motorsports. Audi's tradition in motorsport began with their former company Auto Union in the 1930s. In the 1990s, Audi found success in the Touring and Super Touring categories of motor racing after success in circuit racing in North America.

### Rallying

Walter Röhrl with his Quattro A2 during the 1984 Rally Portugal.

In 1980, Audi released the Quattro, a four-wheel drive (4WD) turbocharged car that went on to win rallies and races worldwide. It is considered one of the most significant rally cars of all time, because it was one of the first to take advantage of the then-recently changed rules which allowed the use of four-wheel drive in competition racing. Many critics doubted the viability of four-wheel drive racers, thinking them to be too heavy and complex, yet the Quattro was to become a successful car. Leading its first rally it went off the road, however the rally world had been served notice 4WD was the future. The Quattro went on to achieve much success in the World Rally Championship. It won the 1983 (Hannu Mikkola) and the 1984 (Stig Blomqvist) drivers' titles,[61] and brought Audi the manufacturers' title in 1982 and 1984.[62]

Audi Quattro S1 driven at the 2007 Rallye Deutschland

In 1984, Audi launched the short-wheelbase Sport Quattro which dominated rally races in Monte Carlo and Sweden, with Audi taking all podium places, but succumbed to problems further into WRC contention. In 1985, after another season mired in mediocre finishes, Walter Röhrl finished the season in his Sport Quattro S1, and helped place Audi second in the manufacturers' points. Audi also received rally honours in the Hong Kong to Beijing rally in that same year. Michèle Mouton, the only female driver to win a round of the World Rally Championship and a driver for Audi, took the Sport Quattro S1, now simply called the "S1", and raced in the Pikes Peak International Hill Climb. The 1439-metre (**unknown operator: u'strong'** ft) climb race pits a driver and car to drive to the summit of the 4302-metre (**unknown operator: u'strong'** ft) Pikes Peak mountain in Colorado, and in 1985, Michèle Mouton set a new record of 11:25.39, and being the first woman to set a Pikes Peak record. In 1986, Audi formally left international rally racing following an accident in Portugal involving driver Joaquim Santos in his Ford RS200. Santos swerved to avoid hitting spectators in the road, and left the track into the crowd of spectators on the side, killing three and injuring 30. Bobby Unser used an Audi in that same year to claim a new record for the Pikes Peak Hill Climb at 11:09.22.

In 1987, Walter Röhrl claimed the title for Audi setting a new Pikes Peak International Hill Climb record of 10:47.85 in his Audi S1, which he had retired from the WRC two years earlier. The Audi S1 employed Audi's time-tested inline-five-cylinder turbocharged engine, with the final version generating 441 kW (**unknown operator: u'strong'** PS; **unknown operator: u'strong'** bhp).[63] The engine was mated to a six-speed gearbox and ran on Audi's famous four-wheel drive system. All of Audi's top drivers drove this car; Hannu Mikkola, Stig Blomqvist, Walter Röhrl and Michèle Mouton. This Audi S1 started the range of Audi 'S' cars, which now represents an increased level of sports-performance equipment within the mainstream Audi model range.

## In the USA

As Audi moved away from rallying and into circuit racing, they chose to move first into America with the Trans-Am in 1988.

In 1989, Audi moved to International Motor Sports Association (IMSA) GTO with the Audi 90, however as they avoided the two major endurance events (Daytona and Sebring) despite winning on a regular basis, they would lose out on the title.

## Touring cars

In 1990, having completed their objective to market cars in North America, Audi returned to Europe, turning first to the Deutsche Tourenwagen Meisterschaft (DTM) series with the Audi V8, and then in 1993, being unwilling to build cars for the new formula, they turned their attention to the fast growing Super Touring series, which are a series of national championships. Audi first entered in the French Supertourisme and Italian Superturismo. In the following year, Audi would switch to the German Super Tourenwagen Cup (known as STW), and then to British Touring Car Championship (BTCC) the year after that.

The Fédération Internationale de l'Automobile (FIA), having difficulty regulating the quattro four-wheel drive system, and the impact it had on the competitors, would eventually ban all four-wheel drive cars from competing in 1998, but by then, Audi switched all their works efforts to sports car racing.

By 2000, Audi would still compete in the US with their RS4 for the SCCA Speed World GT Challenge, through dealer/team Champion Racing competing against Corvettes, Vipers, and smaller BMWs (where it is one of the few series to permit 4WD cars). In 2003, Champion Racing entered an RS6. Once again, the quattro four-wheel drive was superior, and Champion Audi won the championship. They returned in 2004 to defend their title, but a newcomer, Cadillac with the new Omega Chassis CTS-V, gave them a run for their money. After four victories in a row, the Audis were sanctioned with several negative changes that deeply affected the car's performance. Namely, added ballast weights, and Champion Audi deciding to go with different tyres, and reducing the boost pressure of the turbocharger.

In 2004, after years of competing with the TT-R in the revitalised DTM series, with privateer team Abt Racing/Christian Abt taking the 2002 title with Laurent Aïello, Audi returned as a full factory effort to touring car racing by entering two factory supported Joest Racing A4 DTM cars.

## Sports car racing

Further information: List of Audi vehicles (Le Mans prototypes)

Audi R10 TDI

Starting in 1999, Audi built the Audi R8R (open-cockpit 'roadster' prototype) and the Audi R8C (closed-cockpit 'coupé' GT-prototype) to compete in sports car racing, including the Le Mans Prototype LMP900 class at the 24 Hours of Le Mans. For the 2000 season, Audi focused mainly on the new Audi R8, due to favourable rules for open-cockpit prototypes. However, most of the competitors (such as BMW, Toyota, Mercedes and Nissan) retired at the end of 1999. The factory-supported Joest Racing team won at Le Mans three times in a row with the Audi R8 (2000–2002), as well as winning every race in the American Le Mans Series in its first year. Audi also sold the car to customer teams such as Champion Racing.

In 2003, two Bentley Speed 8s, with engines designed by Audi, and driven by Joest drivers *loaned* to the fellow Volkswagen Group company, competed in the GTP class, and finished the race in the top two positions, while the Champion Racing R8 finished third overall, and first in the LMP900 class. Audi returned to the winner's podium at the 2004 race, with the top three finishers all driving R8s: Audi Sport Japan Team Goh finished first, Audi Sport UK

Veloqx second, and Champion Racing third.

At the 2005 24 Hours of Le Mans, Champion Racing entered two R8s, along with an R8 from the Audi PlayStation Team Oreca. The R8s (which were built to old LMP900 regulations) received a narrower air inlet restrictor, reducing power, and an additional 50 kg (**unknown operator: u'strong'** lb) of weight compared to the newer LMP1 chassis. On average, the R8s were about 2–3 seconds off pace compared to the Pescarolo–Judd. But with a team of excellent drivers and experience, both Champion R8s were able to take first and third, while the Oreca team took fourth. The Champion team was also the first American team to win Le Mans since the Gulf Ford GTs in 1967. This also ends the long era of the R8; however, its replacement for 2006, called the Audi R10 TDI, was unveiled on 13 December 2005.

The R10 TDI employed many new and innovative features, the most notable being the twin-turbocharged direct injection diesel engine. It was first raced in the 2006 12 Hours of Sebring as a race-test in preparation for the 2006 24 Hours of Le Mans, which it later went on to win. Audi has been on the forefront of sports car racing, claiming a historic win in the first ever diesel sports car at 12 Hours of Sebring (the car was developed with a Diesel engine due to ACO regulations that favor diesel engines). As well as winning the 24 Hours of Le Mans in 2006 making history, the R10 TDI has also shown its capabilities by beating the Peugeot 908 HDi FAP in 2007, and beating Peugeot again in 2008, (however Peugeot won the 24h in 2009) and, in a podium clean-sweep by proving its reliability throughout the race (compared to all four 908 entries retired before the end of the race) while breaking a new distance record (set way back by the Porsche 917K of Martini Racing in 1971), in 2010 with the R15 TDI Plus.[64]

Audi's sports car racing success would continue with the Audi R18's victory at the 2011 24 Hours of Le Mans. Audi Sport Team Joest's Benoît Tréluyer earned Audi their first pole position in five years while the team's sister car locked out the front row.[65] Early accidents eliminated two of Audi's three entries, but the sole remaining Audi R18 TDI of Tréluyer, Marcel Fässler, and André Lotterer held off the trio of Peugeot 908s to claim victory by a margin of 13.8 seconds.

# Marketing

## Branding

The Audi emblem is four overlapping rings that represent the four marques of Auto Union. The Audi emblem symbolises the amalgamation of Audi with DKW, Horch and Wanderer: the first ring from the left represents Audi, the second represents DKW, third is Horch, and the fourth and last ring Wanderer.[66] [67] Its similarity to the Olympic rings caused the International Olympic Committee to sue Audi in International Trademark Court in 1995.[68]

The old logo used by Audi until 2009

As part of Audi's centennial celebration in 2009, the company updated the logo, changing the font to left-aligned Audi Type, and altering the shading for the overlapping rings.[69] The revised logo was designed by Rayan Abdullah.[70]

### Slogans

Audi's corporate tagline is *Vorsprung durch Technik*, meaning *"Progress through Technology"*.[71] The German-language tagline is used in many European countries, including the United Kingdom, and in other markets, such as Latin America, Oceania and parts of Asia including Japan. A few years ago, the North American tagline was *"Innovation through technology"*, but in Canada the German tagline *Vorsprung durch Technik* was used in advertising. More recently, however, Audi has updated the tagline to "Truth in Engineering" in the U.S.[72]

### Typography

Audi Sans (based on Univers Extended) was originally created in 1997 by Ole Schäfer for MetaDesign. MetaDesign was later commissioned for a new corporate typeface called Audi Type, designed by Paul van der Laan and Pieter van Rosmalen of Bold Monday. The font began to appear in Audi's 2009 products and marketing materials.[73]

## Sponsorships

Audi is a strong partner of different kinds of sports. In association football, long partnerships exist between Audi and various clubs like Manchester United, FC Bayern Munich, Real Madrid CF, FC Barcelona, AC Milan, Ajax Amsterdam and Perspolis F.C.. Audi also sponsors winter sports: The Audi FIS Alpine Ski World Cup is named after the company. Additionally, Audi supports the German Ski Association (DSV) as well as the alpine skiing national teams of Switzerland, Sweden, Finland, France, Liechtenstein, Italy, Austria and the US. For almost two decades Audi fosters golf sport: for example with the Audi quattro Cup and the HypoVereinsbank Ladies German Open presented by Audi. In sailing, Audi is engaged in the Medcup regatta and supports the team Luna Rossa during the Louis Vuitton Pacific Series and also is the primary sponsor of the Melges 20 sailboat. Further, Audi sponsors the regional teams ERC Ingolstadt (hockey) and FC Ingolstadt 04 (soccer).[74] In 2009, the year of Audis 100th anniversary, the company organises the Audi Cup for the first time. In a two-day-tournament, the teams of FC Bayern Munich, AC Milan, Manchester United F.C. and CA Boca Juniors will compete against each other.[75] Audi also sponsor the New York Yankees as well. In October 2010 they agreed to a three sponsorship year-deal with Everton.[76]

## Multitronic campaign

In 2001, Audi promoted the new multitronic continuously variable transmission with television commercials throughout Europe, featuring an impersonator of musician and actor Elvis Presley.[77] [78] A prototypical dashboard figure – later named "Wackel-Elvis" ("Wobble Elvis" or "Wobbly Elvis") – appeared in the commercials to demonstrate the smooth ride in an Audi equipped with the multitronic transmission. The dashboard figure was originally intended for use in the commercials only, but after they aired the demand for Wackel-Elvis fans grew among fans and the figure was mass-produced in China and marketed by Audi in their factory outlet store.[79]

Audi Centre Sydney, Zetland, New South Wales, Australia.

## Audi TDI

As part of Audi's attempt to promote its Diesel technology in 2009, the company began Audi Mileage Marathon. The driving tour featured a fleet of 23 Audi TDI vehicles from 4 models (Audi Q7 3.0 TDI, Audi Q5 3.0 TDI, Audi A4 3.0 TDI, Audi A3 Sportback 2.0 TDI with S tronic transmission) travelling across the American continent from New York to Los Angeles, passing major cities like Chicago, Dallas and Las Vegas during the 13 daily stages, as well as natural wonders including the Rocky Mountains, Death Valley and the Grand Canyon.[80]

## Audi e-tron

The next phase of technology Audi is developing is the e-tron electric drive powertrain system. They have shown several concept cars as of March 2010, each with different levels of size and performance. The original e-tron concept shown at the 2009 Frankfurt motor show is based on the platform of the R8 and has been scheduled for limited production. Power is provided by electric motors at all four wheels. The second concept was shown at the 2010 Detroit Motor Show. Power is provided by two electric motors at the rear axle. This concept is also considered to be the direction for a future mid-engined gas-powered 2-seat performance coupe. The Audi A1 e-tron concept, based on the Audi A1 production model, is a hybrid vehicle with a range extending Wankel rotary engine to provide power after the initial charge of the battery is depleted. It is the only concept of the three to have range extending capability. The car is powered through the front wheels, always using electric power. It is all set to be displayed at the Auto Expo 2012 in New Delhi, India, from the 5 January. Powered by a 1.4 litre engine, and can cover a distance up to 54 km s on a single charge.

## In video games

Audi R8

In PlayStation Home, the PlayStation 3's online community-based service, Audi has supported Home by releasing a dedicated Home space in the European version of Home. Audi is the first carmaker to develop a space for Home. On 17 December 2009, Audi released the Audi Space as two spaces; the Audi Home Terminal and the Audi Vertical Run.[81] The Audi Home Terminal features an Audi TV channel delivering video content, an Internet Browser feature, and a view of a city. The Audi Vertical Run is where users can access the mini-game Vertical Run, a futuristic mini-game featuring Audi's e-tron concept. Players collect energy and race for the highest possible speeds and the fastest players earn a place in the Audi apartments located in a large tower in the centre of the Audi Space. In both the Home Terminal and Vertical Run spaces, there are teleports where users can teleport back and forth between the two spaces. Audi has stated that additional content will be added in 2010.[82]

"Most young people gain their first driving experience from video games", explains Kai Mensing, who is responsible for video games and virtual worlds in Online Marketing at Audi. "With the Audi Space, we can bring this target group into contact with our brand in a highly emotion-packed and interactive environment, and demonstrate our 'Vorsprung durch Technik' with the virtual e-tron race." Kai Mensing's most recent statement on the Audi Space was that "As a brand, Audi sees great value and potential in the experiences and level of interaction the world of gaming affords. PlayStation Home gives Audi the perfect environment to connect with gamers in an unexpected way and we made sure to push the platform to its limits with the new Audi Space."

# References


[1] http://deutsche-boerse.com/dbag/dispatch/en/isg/gdb_navigation/home?module=InOverview_Equi&wp=DE0006757008&foldertype=_Equi&wplist=DE0006757008&active=overview&wpbpl=

[2] "1909 – Audi Automobilwerke is established in Zwickau" (http://www.audiusa.com/audi/us/en2/Company/history/History/chronicle_1899_1914.html). Audi of America. . Retrieved 5 April 2009.

[3] "AUDI AG Finances 2009" (http://www.audiusa.com/etc/medialib/ngw/company/investor_relations/pdf/finanzberichte/geschaeftsberichte4.Par.0002.File.pdf/audi_ag_lagebericht.pdf). Audi AG. pp. 22, 40, 41. . Retrieved 24 January 2011.

[4] http://www.audi.com/

[5] "Audi selected as the most creative premium brand in HORIZONT ranking, retrieved 29 January 2011" (http://www.oneighturbo.com/audi/audi-selected-as-the-most-creative-premium-brand-in-horizont-ranking/). Oneighturbo.com. 10 December 2009. . Retrieved 12 March 2011.

[6] "List of Shareholdings in accordance with sections 285 and 313 of the HGB of Volkswagen AG and the Volkswagen Group as of December 31, 2009" (http://www.volkswagenag.com/vwag/vwcorp/info_center/en/publications/2010/03/Anteilsbesitz.-bin.acq/qual-BinaryStorageItem.Single.File/Anteilsbesitz_2009_e.pdf). *www.volkswagenag.com*. Volkswagen AG. 31 December 2009. . Retrieved 11 December 2010.

[7] "motorauthority.com" (http://www.motorauthority.com/blog/1028368_porsche-to-buy-audi-lamborghini-bugatti-and-bentley). . Retrieved 15 January 2010.

[8] Volkswagen's branding plan takes shape (http://www.autonews.com/apps/pbcs.dll/article?AID=/20011203/ANE/112030837)

[9] "Volkswagen Group – Business lines and markets" (http://www.volkswagenag.com/vwag/gb2006/content/en/Divisions/Business_Lines_and_Markets.html). Volkswagenag.com. . Retrieved 9 January 2011.

[10] "Volkswagen Group – Brands and business fields" (http://www.volkswagenag.com/vwag/gb2007/content/en/divisions/business_lines_and_markets.html). Volkswagenag.com. . Retrieved 9 January 2011.

[11] "Brand family tree" (http://www.audiusa.com/us/brand/en/about/main/history/brand_family_tree/brand_family_tree.html). audiusa com. . Retrieved 15 August 2010.

[12] Audi website http://www.audiusa.com/us/brand/en/about/main/history.html

[13] Audi AG motion picture 1994: "The Silver Arrows from Zwickau", running time approx. 49 mins.

[14] August Horch: "Ich baute Autos – Vom Schmiedelehrling zum Autoindustriellen", Schützen-Verlag Berlin 1937

[15] *A History of Progress – Chronicle of the Audi AG* (http://www.bentleypublishers.com/product.htm?code=gahp). Audi AG, Public Relations. 1996. p. 30. ISBN 0837603846. .

[16] Baldwin, Nick; Laban, Brian (1987). *The World guide to automobile manufacturers*. Facts on File Publications. p. 43. ISBN 9780816018444.

[17] Audi website http://www.audi.com/audi/com/en2/about_audi_ag/history/chronicle/chronicle_1915_1929.html

[18] Audi website http://www.audi.com/audi/com/en2/about_audi_ag/history/chronicle/chronicle_1930_1944.html

[19] Audi website http://www.audi.com/audi/com/en2/about_audi_ag/history/chronicle/chronicle_1945_1959.html

[20] "VEB Automobilwerk Zwickau" (http://de.wikipedia.org/wiki/Automobilwerk_Zwickau) (in German). . Retrieved 21 September 2008.

[21] "Audi Worldwide: Home" (http://www.audi.com/audi/com/en2/about_audi_ag/history/chronicle/chronicle_1945_1959.html). Audi.com. 15 April 2009. . Retrieved 27 April 2009.

[22] Bulmer (Ed), Charles (24 July 1971). "The Audi-NSU Affair". *Motor*: 21.

[23] Turner, Philip (24 July 1971). "Turner's Travel [to Wolfsburg]". *Motor*: 28–30.

[24] "Quattro, The early years" (http://www.worldcarfans.com/news.cfm/newsID/2050225.007/country/gcf/audi/audi-quattro-the-early-years). . Retrieved 2 November 2006.

[25] Cremer, Andreas; Lavell, Tom (4 February 2010). "Audi 1980s Scare May Mean Lost Generation for Toyota" (http://www.businessweek.com/news/2010-02-04/audi-s-1980s-scare-may-mean-lost-generation-for-toyota-sales.html). *Business Week*. . Retrieved 15 August 2010.

[26] Holusha, John (24 July 1988). "A Hard Sell for Audi" (http://www.nytimes.com/1988/07/24/business/a-hard-sell-for-audi.html?scp=14&sq=audi sudden acceleration&st=cse). *The New York Times*. . Retrieved 25 April 2010.

[27] Niedermeyer, Paul (7 March 2010). "The Audi 5000 Intended Unintended acceleration Debacle" (http://www.thetruthaboutcars.com/the-best-of-ttac-the-audi-5000-intended-unintended-acceleration-debacle/). The Truth About Cars. . Retrieved 15 August 2010.

[28] Huber, Peter (January 1990). "Manufacturing the Audi Scare" (http://www.manhattan-institute.org/html/cjm_18.htm). Manhattan Institute for Policy Research. . Retrieved 15 August 2010.

[29] Gossett, Sherrie (13 May 2005). "The CBS "Cold Case" Files" (http://www.aim.org/media-monitor/the-cbs-cold-case-files/). Accuracy in Media. . Retrieved 15 August 2010.

[30] Yates, Brock (16 April 1989). "Unfair at Any Speed" (http://www.highbeam.com/doc/1P2-1185635.html). *The Washington Post*. highbeam com. . Retrieved 15 August 2010.

[31] "Audi Increases Warranty Plan" (http://www.nytimes.com/1988/07/27/business/company-news-audi-increases-warranty-plan.html?scp). *The New York Times*. 27 July 1988. . Retrieved 15 August 2010.

[32] "China: Audi sells a million | beyondbrics | News and views on emerging markets from the Financial Times – FT.com" (http://blogs.ft.com/beyond-brics/2010/10/20/audis-millionth-car-sale-shows-china-still-tops/). Blogs.ft.com. 20 October 2010. . Retrieved 9 January 2011.

[33] Canada. "Audi's A game" (http://www.theglobeandmail.com/globe-drive/new-cars/auto-news/audis-a-game/article1377213/). *The Globe and Mail*. . Retrieved 12 March 2011.

[34] "Audi Worldwide > Company > Production plants > Aurangabad" (http://www.audi.com/com/brand/en/company/production_plants/aurangabad.html#source=http://www.audi.com/com/brand/en/company/production_plants.html&container=page). Audi.com. 27 August 2010. . Retrieved 12 March 2011.

[35] Audi website http://www.audiusa.com/audi/us/en2/tools/glossary/chassis_body/corrosion_protection.html

[36] Audi of America - http://www.audiusa.com.  "Audi of America > Glossary > Chassis & Body > Galvanised body" (http://www.audiusa.com/audi/us/en2/tools/glossary/chassis_body/galvanised_body.html). Audiusa.com. . Retrieved 27 April 2009.

[37] "2007 Audi S8 - Long-Term Road Test - Page 4" (http://www.caranddriver.com/reviews/car/09q1/2007_audi_s8-long-term_road_test/sidebar_page_3). Caranddriver.com. . Retrieved 2011-11-18.

[38] Peterson, Thane (5 October 2010). "Review: 2011 Audi A8" (http://www.businessweek.com/lifestyle/content/oct2010/bw2010104_140585.htm). *BusinessWeek*. . Retrieved 2 February 2011.

[39] Wan, Mark. "Volkswagen's VR6 and W-engines – W12" (http://www.autozine.org/technical_school/engine/tech_engine_packaging.htm#W12). *AutoZine.org*. . Retrieved 31 December 2009.

[40] "2005 Audi A8 Review and Photos" (http://www.newcartestdrive.com/review-intro.cfm?Vehicle=2005_Audi_A8&ReviewID=2934). New Car Test Drive. . Retrieved 2011-11-18.

[41] "2010 Audi R8 LED Headlights" (http://www.caranddriver.com/features/2010-audi-r8-led-headlights). Car and Driver. . Retrieved 28 January 2012.

[42] "New Audi S6" (http://www.audi.co.uk/new-cars/a6/s6-saloon.html). www.audi.co.uk. . Retrieved 2011-09-17.

[43] "New Audi S7" (http://www.audi.co.uk/new-cars/a7/s7-sportback.html). www.audi.co.uk. . Retrieved 2011-09-17.

[44] "New Audi S8" (http://www.audi.co.uk/new-cars/a8/s8.html). www.audi.co.uk. . Retrieved 2011-09-17.

[45] Error - LexisNexis® Publisher (http://www6.lexisnexis.com/publisher/EndUser?Action=UserDisplayFullDocument&orgId=101846&topicId=103840033&docId=l:788745336)

[46] "Audi A1 Sportback concept" (http://www.nextconceptcars.com/concept-cars/audi-a1-sportback-concept/). Next Concept Cars. 2 October 2008. . Retrieved 27 April 2009.

[47] Korzeniewski, Jeremy (2 October 2008). "Audi unveils A4 TDI concept e" (http://www.autobloggreen.com/2008/10/02/audi-unveils-a4-tdi-concept-e/). Autobloggreen.com. . Retrieved 27 April 2009.

[48] "Audi e-Tron Electric Supercar Concept Unveiled" (http://www.audisite.com/1008/audi-e-tron-electric-concept-unveiled/). Audisite.com. .

[49] Volkswagen AG Annual Report 1999 (Including 1998) [Page 50] http://www.volkswagenag.com/vwag/vwcorp/info_center/en/publications/2000/03/Annual_Report_1999.-bin.acq/qual-BinaryStorageItem.Single.File/VW_AR_1999_e.pdf

[50] Volkswagen AG Annual Report 2000 [Page 53] http://www.volkswagenag.com/vwag/vwcorp/info_center/en/publications/2001/03/Annual_Report_2000.-bin.acq/qual-BinaryStorageItem.Single.File/VW_GB_2000_e.pdf

[51] Volkswagen AG Annual Report 2001 [Page 41] http://www.volkswagenag.com/vwag/vwcorp/info_center/en/publications/2002/03/annual_report_2001.-bin.acq/qual-BinaryStorageItem.Single.File/20020312_GB_2001_e.pdf

[52] Volkswagen AG Annual Report 2002 [Page 77] http://www.volkswagenag.com/vwag/vwcorp/info_center/en/publications/2003/03/annual_report_2002.-bin.acq/qual-BinaryStorageItem.Single.File/vw_gb_2002_en.pdf

[53] Volkswagen AG Annual Report 2003 [Page 97] http://www.volkswagenag.com/vwag/vwcorp/info_center/en/publications/2004/03/annual_report_2003.-bin.acq/qual-BinaryStorageItem.Single.File/20040309_gb2003_e.pdf

[54] Volkswagen AG Annual Report 2004 [Page 91] http://www.volkswagenag.com/vwag/vwcorp/info_center/en/publications/2005/03/annual_report_2004.-bin.acq/qual-BinaryStorageItem.Single.File/Gbericht_2004_en.pdf

[55] Volkswagen AG Annual Report 2005 [Page 41] http://www.volkswagenag.com/vwag/vwcorp/info_center/en/publications/2006/03/Volkswagen_AG__Annual_Report_2005.-bin.acq/qual-BinaryStorageItem.Single.File/gesch%C3%A4ftsbericht_2005_english.pdf

[56] Volkswagen AG Annual Report 2006 [Page 45] http://www.volkswagenag.com/vwag/vwcorp/info_center/en/publications/2007/03/Geschaeftsbericht_2006.-bin.acq/qual-BinaryStorageItem.Single.File/Annual_Report_2006_e.pdf

[57] Volkswagen AG Annual Report 2007 [Page 83] http://www.volkswagenag.com/vwag/vwcorp/info_center/en/publications/2008/03/Annual_Report_2007.-bin.acq/qual-BinaryStorageItem.Single.File/VW_AG_GB_2007_en.pdf

[58] Volkswagen AG Annual Report 2008 [Page 83] http://www.volkswagenag.com/vwag/vwcorp/info_center/en/publications/2009/03/GB_2008.-bin.acq/qual-BinaryStorageItem.Single.File/Y_2008_e.pdf

[59] Volkswagen AG Annual Report 2009 [Page 93] http://www.volkswagenag.com/vwag/vwcorp/info_center/en/publications/2010/03/Annual_Report_2009.-bin.acq/qual-BinaryStorageItem.Single.File/Y_2009_e.pdf

[60] Volkswagen AG Annual Report 2010 [Page 111] http://www.volkswagenag.com/vwag/vwcorp/info_center/en/publications/2011/03/Volkswagen_AG_Geschaeftsbericht_2010.-bin.acq/qual-BinaryStorageItem.Single.File/GB_2010_e.pdf

[61] "World Rally Championship for Drivers – Champions" (http://www.rallybase.nl/index.php?type=championlist&subchamptype=wcd). *RallyBase.nl*. . Retrieved 30 August 2008.

[62] "World Rally Championship for Manufacturers – Champions" (http://www.rallybase.nl/index.php?type=championlist&subchamptype=wrc). *RallyBase.nl*. . Retrieved 30 August 2008.

[63] "25 Years of Audi Quattro" (http://www.audiusanews.com/newsrelease.do?id=211). *Audi AG*. Audi Of America. 22 February 2005. . Retrieved 31 March 2009.

[64] "Audi achieves record victory at Le Mans with new technology" (http://www.joest-racing.de/web/news/news.php?id=541&sprache=e). *joest-racing.de; Audi Sport*. Joest Racing. 13 June 2010. . Retrieved 16 June 2010.
[65] "Audi back on front row for Le Mans 24 Hours" (http://www.google.com/hostednews/afp/article/ALeqM5ikxFTgOVDE0DXKGgxuzUurSVk4dw?docId=CNG.4facb82ec8689e5a037eb4e7286eff0b.3b1). AFP. 10 June 2011. .
[66] Car Logo. "Audi Logo" (http://www.carlogo.info/audi-audi-logo_blog-bid-29.htm). . Retrieved 10 September 2007.
[67] "History of the Four Rings-Part 1-Audi Auto Union" (http://www.seriouswheels.com/art-four-rings-1.htm). Seriouswheels.com. . Retrieved 27 April 2009.
[68] Alina Dumitrache. "Audi Reveals Updated Logo" (http://www.autoevolution.com/news/audi-reveals-updated-logo-10315.html). autoevolution.com. . Retrieved 27 August 2009.
[69] "Audi Unveils Updated Logo Following Centennial Celebration" (http://wot.motortrend.com/6546262/marketing/audi-unveils-updated-logo-following-centennial-celebration/index.html). Wot.motortrend.com. . Retrieved 2 August 2010.
[70] "mosul-network.org" (http://www.mosul-network.org/index.php?do=article&id=18707). mosul-network.org. 2011-01-20. . Retrieved 2011-11-18.
[71] "Eco-Culture". *Audi magazine* (3/08): 19.
[72] Lavrinc, Damon. Audi planning TT and R8 lightweight "Sport" models (http://www.autoblog.com/2008/04/08/audi-planning-tt-and-r8-lightweight-sport-models/). *Autoblog*. Retrieved 24 April 2010.
[73] "Bold Monday: Audi Type" (http://www.boldmonday.com/en/audi). boldmonday.com. . Retrieved 6 October 2010.
[74] "Audi Worldwide > Experience > Sponsoring > Sport" (http://www.audi.com/com/brand/en/experience/sponsoring/sportsponsoring.html). Audi.com. 11 June 2009. . Retrieved 7 July 2009.
[75] "FC Bayern" (http://www.fcbayern.t-home.de/en/news/news/2009/18772.php). Fcbayern.t-home.de. 4 March 2009. . Retrieved 7 July 2009.
[76] "Blues agree deal with Audi" (http://www.evertonfc.com/news/archive/2010/10/14/blues-agree-deal-with-audi). evertonfc.com. 14 October 2010. . Retrieved 16 March 2011.
[77] "". "Audi Wackel-Elvis commercial (2001, British version)" (http://www.youtube.com/watch?v=YQo95oI4nXY). *Youtube*. . Retrieved 2 August 2010.
[78] "". "Audi Wackel-Elvis commercial (2001, German version)" (http://www.youtube.com/watch?v=b1czNu9pTzM). *Youtube*. . Retrieved 2 August 2010.
[79] "Fans Waiting in Line for Release of Wackel-Elvis" (http://www.welt.de/print-welt/article456186/Wackel_Elvis_von_Audi_Fans_stehen_Schlange.html) (in (German)). *Die Welt*. 11 June 2001. . Retrieved 2 August 2010.
[80] Abuelsamid, Sam (8 September 2008). "Audi to kick off diesel push with cross country Mileage Marathon" (http://green.autoblog.com/2008/09/08/audi-to-kick-off-diesel-push-with-cross-country-mileage-marathon/). Green.autoblog.com. . Retrieved 2 August 2010.
[81] Fernandez, Joe (10 December 2009). "Audi to launch PlayStation download game" (http://www.marketingweek.co.uk/audi-to-launch-playstation-download-game/3007770.article). Marketing Week. . Retrieved 10 December 2009.
[82] "Audi creates virtual Audi Space within PlayStation Home" (http://reviews.cnet.com/8301-13746_7-10358137-48.html). CNET Reviews. .

It is all set to be displayed at the Auto Expo 2012 in New Delhi, India, from the 5 January.

# External links

- Global corporate portal (http://www.audi.com/)
- Audi Service (http://www.servicingstop.co.uk/audi_service.html)

# Wrigley_Company

| WRIGLEY | |
|---|---|
| **Type** | Subsidiary |
| **Industry** | Food processing |
| **Founded** | Chicago, Illinois, U.S. (1891) |
| **Founder(s)** | William Wrigley, Jr. |
| **Headquarters** | Wrigley Building<br>Near North Side, Chicago, U.S. |
| **Key people** | William Wrigley, Jr. II, Chairman<br>Martin Radvan, CEO |
| **Products** | Juicy Fruit, Wrigley's Spearmint, Doublemint, more... |
| **Revenue** | ▲ $5.389 billion (2007) |
| **Net income** | ▲ $961.9 million (2009) |
| **Employees** | 16,000 |
| **Parent** | Mars, Incorporated (2008–present) |
| **Website** | www.wrigley.com [1] |

The **William Wrigley Jr. Company** is a company headquartered in the Wrigley Building in Near North Side, Chicago, Illinois.[2] The company was founded on April 1, 1891, originally selling products such as soap and baking powder. In 1892, William Wrigley, Jr., the company's founder, began packaging chewing gum with each can of baking powder. The chewing gum eventually became more popular than the baking powder itself and Wrigley's reoriented the company to produce the popular chewing gum.

Wrigley Building

The company currently sells its products in more than 180 countries and maintains 14 factories in various countries, including the United States, Mexico, Australia, the United Kingdom, Canada, Spain, New Zealand, the Philippines, France, Kenya, Taiwan, China, India, Poland, and Russia.

In 2004, Wrigley purchased the Life Savers and Altoids businesses from Kraft Foods for US$1.48 billion. On January 23, 2007, Wrigley signed a purchase agreement to acquire an 80 percent initial interest in A. Korkunov for $300 million with the remaining 20 percent to be acquired over time. On April 28, 2008, it was announced that Mars, Inc would acquire Wrigley for approximately $23 billion.[3] Financing for the transaction was provided by Berkshire Hathaway, Goldman Sachs and JPMorgan. Berkshire Hathaway holds a minority equity investment in the Wrigley subsidiary.

The corporate headquarters, the Wrigley Building, is one of the more well-known landmarks on the Magnificent Mile in Chicago.

## Corporate leadership

On October 23, 2006, William D. Perez succeeded Bill Wrigley Jr. as CEO. He is the first person outside the Wrigley family to head the 120-year-old company. Bill Wrigley Jr. retained the post of executive chairman.

CEO line:

- 1891–1932 - William Wrigley Jr. (1861–1932)
- 1932–1961 - Philip K. Wrigley (1894–1977)
- 1961–1999 - William Wrigley III (1933–1999)
- 1999–2006 - William Wrigley, Jr. II (1963–)
- 2006–2008 - William Perez
- 2008–2011 - Dushan "Duke" Petrovich
- 2011 - Present - Martin Radvan

## Recent changes in gum

In June 1974, a Marsh supermarket in Troy, Ohio installed the first bar code scanning equipment. The first product to be scanned using a UPC bar code was a 10 pack of Wrigley's Juicy Fruit gum. This pack of gum is now on display at the Smithsonian Institution's National Museum of American History.

In some countries, xylitol is used to sweeten gum, instead of aspartame. By avoiding sugar, the chance of tooth decay is lowered, since the sugar otherwise used may turn into acid after chewing the gum. It is also claimed that in chewing, it may help to remove food residues. Xylitol based products are allowed by the U.S. Food and Drug Administration to make the medical claim that they do not promote dental cavities.[4]

## Subsidiaries

- The Wrigley Company Limited
- Amurol Confections Company
- Northwestern Flavors, LLC

## Brands

### Gum (United States)

- Juicy Fruit (1893)
- Spearmint (1893)
- Doublemint (1914)
- Freedent (1975)
- Big Red (1975)
- Extra (1984)
- Winterfresh (1994)
- Surpass (2001)
- Orbit (reintroduced 2001)
    - Orbit White
- Eclipse (2001)
- 5 (2007)

A newspaper ad from 1920 for 3 types of Wrigley's gum.

### Gum (Canada)

- 5
- Doublemint
- Excel
- Excel Mist
- Excel White
- Extra
- Extra Professional
- Extra Professional White
- Hubba Bubba
- Juicy Fruit
- Life Savers
- Skittles
- Starburst

### Gum (made by The Wrigley Company Ltd., Estover, Plymouth, UK)

- Active8 (discontinued)
- Airwaves
- Big Red (discontinued in 1998)
- Doublemint
- Extra
- 5
- Freedent (discontinued)
- Ice Skating - Later re-branded as Buy one pack of Ice White and get ice skating lessons free
- Juicy Fruit
- Orbit
  - Orbit Professional
  - Orbit Complete
- P.K
- Wrigley's Spearmint
- Fergie's juicy gum

### Products

- Altoids
- Big League Chew (until November 2010)
- Bubble Tape
- Cool Air
- Eclipse
- Excel
- Hubba Fergie
- Spec Savers
  - Gummi Savers
  - Life Saver Minis
  - Life Saver Fusions
  - Creme Savers
- Lucas (Spicy)
- Skwinkles (Spicy)
- Starburst
- Skittles
- Lockets
- Tunes

## Recycling

TerraCycle repurposes waste packaging from Mars Incorporated and Wrigley's operations into products ranging from cell phone holders and laptop sleeves to messenger bags.

## See also

- Wrigley Building
- Wrigley Field—Chicago
- Wrigley Field—Los Angeles
- Wrigley Institute for Environmental Studies—Catalina Island
- Wrigley Rooftops
- Wrigley Square
- Wrigleyville

## References

[1] http://www.wrigley.com/
[2] " Contact Us (http://www.wrigley.com/global/contact-us.aspx)." Wrigley Company. Retrieved on June 26, 2010.
[3] Karnitschnig, Matthew; Berman, Dennis K. (2008-04-27). "Mars, Buffett Team Up in Wrigley Bid" (http://online.wsj.com/article/SB120935192240148985.html?mod=hps_us_whats_news#printMode). Wall Street Journal. . Retrieved 2008-04-27.
[4] http://www.accessdata.fda.gov/scripts/cdrh/cfdocs/cfcfr/CFRSearch.cfm?fr=101.80

## External links

- Official site (http://www.wrigley.com)
- Wrigley Candystand (http://www.candystand.com/index.do)
- Wrigley's press release (http://www.corporate-ir.net/ireye/ir_site.zhtml?ticker=WWY&script=410&layout=0&item_id=952887)

### Data

- Yahoo! — Company Profile (http://biz.yahoo.com/ic/11/11648.html)

# Personal_computer_hardware

**Personal computer hardware** are the component **devices** that are the building blocks of personal computers. These are typically installed into a computer case, or attached to it by a cable or through a port. In the latter case, they are also referred to as peripherals.

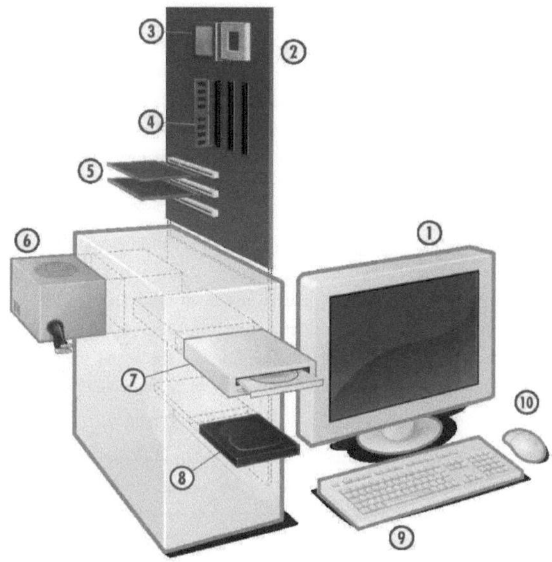

Hardware of a modern personal computer 1. Computer monitorMonitor 2. Motherboard3. Central Processing UnitCPU4. Random-access memoryRAM5. Expansion cards6. Power supply unit (computer)Power supply7. Optical disc drive8. Hard disk drive9. Computer keyboardKeyboard10. Computer MouseMouse

## Computer Case

A computer case (also known as a computer chassis, cabinet, box, tower, enclosure, housing, system unit or simply case) is the enclosure that contains most of the components of a computer (usually excluding the display, keyboard and mouse). A computer case is sometimes incorrectly referred to metonymously as a CPU referring to a component housed within the case. CPU was a common term in the earlier days of home computers, when peripherals other than the motherboard were usually housed in their own separate cases.

## Power supply

A power supply unit (PSU) converts alternating current (AC) electric power to low-voltage DC power for the internal components of the computer. Some power supplies have a switch to change between 230 V and 115 V. Other models have automatic sensors that switch input voltage automatically, or are able to accept any voltage between those limits. Power supply units used in computers are nearly always switch mode power supplies (SMPS). The SMPS provides regulated direct current power at the several voltages required by the motherboard and accessories such as disk drives and cooling fans.

Inside a custom-built computer: the power supply at the bottom has its own cooling fan.

## Motherboard

The motherboard is the main component inside the case. It is a large rectangular board with integrated circuitry that connects the other parts of the computer including the CPU, the RAM, the disk drives (CD, DVD, hard disk, or any others) as well as any peripherals connected via the ports or the expansion slots.

Components directly attached to the motherboard include:

- The **CPU** (Central Processing Unit) performs most of the calculations which enable a computer to function, and is sometimes referred to as the "brain" of the computer. It is usually cooled by a heat sink and fan. Most newer CPUs include an on-die Graphics Processing Unit (GPU).
- The **Chipset**, which includes the north bridge, mediates communication between the CPU and the other components of the system, including main memory.
- The **Random-Access Memory** (RAM) stores the code and data that are being actively accessed by the CPU.
- The **Read-Only Memory** (ROM) stores the BIOS that runs when the computer is powered on or otherwise begins execution, a process known as Bootstrapping, or "booting" or "booting up". The **BIOS** (Basic Input Output System) includes boot firmware and power management firmware. Newer motherboards use Unified Extensible Firmware Interface (UEFI) instead of BIOS.
- **Buses** connect the CPU to various internal components and to expansion cards for graphics and sound.
  - **Current**
    - PCI Express: for expansion cards such as graphics, sound, network interfaces, TV tuners, etc.
    - PCI: for other expansion cards.
    - SATA: for disk drives.
  - **Obsolete**
    - AGP: superseded by PCI Express.
    - ATA
    - VLB: VESA Local Bus, superseded by AGP.
    - EISA
    - Micro Channel architecture
    - ISA: expansion card slot format obsolete in PCs, but still used in industrial computers.
- **Ports** for external peripherals. These ports may be controlled directly by the south bridge I/O controller or provided by expansion cards attached to the motherboard.
  - USB
  - Memory Card
  - FireWire

- eSATA
- SCSI

## Expansion Cards

The expansion card (also expansion board, adapter card or accessory card) in computing is a printed circuit board that can be inserted into an expansion slot of a computer motherboard or backplane to add functionality to a computer system via the expansion bus.

An example of an expansion card is a sound card that enables the computer to output sound to audio devices, as well as accept input from a microphone. Most modern computers have hardware support for sound integrated in the motherboard chipset but some users prefer to install a separate sound card as an upgrade. Most sound cards, either built-in or added, have surround sound capabilities and 3-D sound effects.

## Secondary Storage Devices

Computer data storage, often called storage or memory, refers to computer components and recording media that retain digital data. Data storage is a core function and fundamental component of computers.

### Fixed Media Devices

- Hard disk drives: a hard disk drive (HDD; also hard drive, hard disk, or disk drive)[2] is a device for storing and retrieving digital information, primarily computer data. It consists of one or more rigid (hence "hard") rapidly rotating discs (often referred to as platters), coated with magnetic material and with magnetic heads arranged to write data to the surfaces and read it from them.
- Solid-state drives: a solid-state drive (SSD), sometimes called a solid-state disk or electronic disk, is a data storage device that uses solid-state memory to store persistent data with the intention of providing access in the same manner of a traditional block I/O hard disk drive. SSDs are distinguished from traditional magnetic disks such as hard disk drives (HDDs) or floppy disk, which are electromechanical devices containing spinning disks and movable read/write heads.
- RAID array controller - a device to manage several internal or external hard disks and optionally some peripherals in order to achieve performance or reliability improvement in what is called a RAID array.

### Removable Media Devices

- Optical Disc Drives for reading from and writing to various kinds of optical media, including Compact Discs such as CD-ROMs, DVDs, DVD-RAMs and Blu-ray Discs. Optical discs are the most common way of transferring digital video, and are popular for data storage as well.
- Floppy disk drives for reading and writing to floppy disks, an outdated storage media consisting of a thin disk of a flexible magnetic storage medium. These were once standard on most computers but are no longer in common use. Floppies are used today mainly for loading device drivers not included with an operating system release (for example, RAID drivers).
- Zip drives, an outdated medium-capacity removable disk storage system, for reading from and writing to Zip disks, was first introduced by Iomega in 1994.
- USB flash drive plug into a USB port and do not require a separate drive. USB flash drive is a typically small, lightweight, removable, and rewritable flash memory data storage device integrated with a USB interface. Capacities vary, from hundreds of megabytes (in the same range as CDs) to tens of gigabytes (surpassing Blu-ray discs but also costing significantly more).
- Memory card readers for reading from and writing to Memory cards, a flash memory data storage device used to store digital information. Memory cards are typically used on mobile devices. They are thinner, smaller and

lighter than USB flash drives. Common types of memory cards are SD and MS.
- Tape drives read and write data on a magnetic tape, and are used for long term storage and backups.

## Input and output peripherals

Input and output devices are typically housed externally to the main computer chassis. The following are either standard or very common to many computer systems.

### Input

Wheel mouse

- Text input devices
  - Keyboard - a device to input text and characters by depressing buttons (referred to as keys or buttons).
- Pointing devices
  - Mouse - a pointing device that detects two dimensional motion relative to its supporting surface.
    - Optical Mouse - uses light (laser technology) to determine mouse motion.
  - Trackball - a pointing device consisting of an exposed protruding ball housed in a socket that detects rotation about two axes.
  - Touchscreen - senses the user pressing directly on the monitor.
- Gaming devices
  - Joystick - a hand-operated pivoted stick whose position is transmitted to the computer.
  - Game pad - a hand held game controller that relies on the digits (especially thumbs) to provide input.
  - Game controller - a specific type of controller specialized for certain gaming purposes.
- Image, Video input devices
  - Image scanner - a device that provides input by analyzing images, printed text, handwriting, or an object.
  - Web cam - a video camera used to provide visual input that can be easily transferred over the internet.
- Audio input devices
  - Microphone - an acoustic sensor that provides input by converting sound into electrical signals.

### Output

- Printer - a device that produces a permanent human-readable text or graphic document.
  - Laser printer
  - Inkjet printer
  - Dot matrix printer
  - Thermal printer
- Computer monitors
- Speakers

## See also

- Computer hardware
- Networking hardware
- Glossary of computer terms

## References

# Dock_Connector

A **dock connector** is a connector used to attach a mobile electronic device simultaneously to multiple external resources. The dock connector will typically carry a variety of signals and power, through a single connector, to simplify the process of docking the mobile device. A dock connector may be embedded in a mechanical fixture used to support or align the mobile device or may be at the end of a cable.

The dock connector was originally associated with laptops, but other mobile devices use the concept.

## Laptop dock connector

Docking connectors for laptop computers are usually embedded into a mechanical device that supports and aligns the laptop and sports various single function ports and a power source that are aggregated into the docking connector. Early docking connectors would carry keyboard, serial, parallel, and video ports from the laptop and supply power to it. Often these ports were available on the laptop itself, giving rise to the term "port replicator" for simpler docks. More complicated docks are usually termed docking stations, but both will have some sort of docking connector. As laptop interfaces evolved, so did the functions carried by the dock connector. The keyboard and serial ports changed to PS/2 connector for the keyboard and mouse. Now, USB is almost universal and DVI is common. eSATA and FireWire ports are also carried through the docking connector.

## Portable media player dock connector

An Apple Dock Connector

Many Portable media players feature a dock connector, such as the Apple dock connector, which is a proprietary 30-pin connector that is common on most models of iPods, and all models of the iPhone and iPad.[1] Originally, the Apple dock connector carried USB, FireWire, some controls and line-level audio outputs. As the iPod evolved, so did the signals in the dock connector. Video was added to the connector. FireWire was phased out of the iPods, which led to a discontinuity in usage of the dock connector. Older implementations of the dock connector may have been using the FireWire power pin to supply power, and cannot charge later iPods, iPhones, or the iPad, requiring an after market adapter to allow charging.

The simplest form of the dock connector today takes the form of a simple cable with the dock connector on one end and a USB connector on the other. This cable only has the USB signals in the dock connector, but is sufficient to synchronize and charge the devices. At the other end of the spectrum, dock connectors are used to completely merge devices into a variety of accessories, both mechanically and electrically.

Because of the popularity of portable media players, many third parties have used dock connector interfaces for accessories. This includes many variations of external speakers, including stereo systems and clock radios. Automotive accessories for the portable media players are numerous and include charging cradles, FM transmitters for playing audio through the car's speakers and a full GPS receiver. There are also original and aftermarket cables with dock connectors for direct integration with the car's audio system and controls.

## See also

- Docking station
- PDMI, a portable device interconnection standard from CEA

## References

[1] iPod Hi-Fi User guide. "Chapter 2, Step 2: Connect your iPod, page 13" (http://manuals.info.apple.com/en/iPod_Hi-Fi_Users_Guide.pdf) (PDF). *Apple.com.* . Retrieved 2007-05-25.

## External links

- Apple iPod, iPhone (2g, 3g), iPad dock connector pinout and interfacing (http://pinouts.ru/PortableDevices/ipod_pinout.shtml)
- (http://www.allpinouts.org/index.php/Apple_iPod,_iPad_and_iPhone_dock)—Apple connector pinout
- (http://nuxx.net/wiki/iPod_Dock_Connector)—Apple connector sources and specifications

# iPod Classic

iPod Classic (6th generation)

| Manufacturer | Apple Inc. |
|---|---|
| Type | Portable media player / Digital audio player |
| Retail availability | October 23, 2001 – present |
| Media | Hard drive<br>From 5–160 GB (currently 160 GB) |
| Operating system | 1.5 (1G, 2G)<br>2.3 (3G)<br>3.1.1 (4G)<br>1.2.1 (4G Color)<br>1.3 (5G)<br>1.1.2 (6G)<br>2.0.4 (6.5G) |
| Power | Lithium polymer battery<br>Lithium ion battery |
| Display | **1G-4G:** 160 x 128 px, 2 in (**unknown operator: u'strong'** mm), monochrome LCD<br>**Color:** 220 x 176 px, 2 in (**unknown operator: u'strong'** mm), color LCD<br>**5G-6G:** 320 x 240 px, 2.5 in (**unknown operator: u'strong'** mm), color LCD |
| Input | **4G-6G:** Click wheel<br>**1G:** Scroll wheel<br>**2G-3G:** Touch wheel |
| Connectivity | **3G-6G:** USB 2.0<br>**1G-4G:** FireWire |
| Related articles | iPod Shuffle<br>iPod Nano<br>iPod Touch<br>iPhone |

The **iPod Classic** (trademarked, marketed, and stylized as **iPod classic** and known before its sixth generation as simply **iPod**) is a portable media player marketed by Apple Inc. The current generation is by far the most capacious iPod, with 160GB of storage.

To date, there have been six generations of the iPod, as well as a spin-off (the iPod Photo) that was later re-integrated into the main iPod line. (Some sources incorrectly refer to the revisions of the sixth generation as a separate "seventh generation."[1] ) All generations use a 1.8-inch (**unknown operator: u'strong'** mm) hard drive for storage. The "classic" suffix was initially introduced when a freelance writer analyzing eBay's used/broken iPod marketplace categorized iPods into different types on May 21, 2006 [2] and was formally introduced with the rollout of the sixth-generation iPod on September 5, 2007 [3] prior to this, all iPod models were simply referred to as iPods.[4] It is available in silver or black replacing the "signature iPod white".

## History

### First generation

A first generation iPod (2001)

Apple introduced the first-generation iPod on October 23, 2001, with the slogan "1,000 songs in your pocket". The first iPod had a black and white LCD (liquid-crystal display) screen and featured a 5 GB hard drive capable of storing 1,000 songs encoded using MP3 and was priced at US$399. Among the iPod's innovations were its small size, achieved using a 1.8" hard drive, whereas its competitors were using 2.5" hard drives at the time, and its easy-to-use navigation, which was controlled using a mechanical scroll wheel (unlike later iPods, which had touch-sensitive scroll wheels), a center select button, and four auxiliary buttons around the wheel. The iPod had a rated battery life of ten hours.

On March 20, 2002, Apple introduced a 10 GB model of the first generation iPod for US$499. vCard compatibility was added, as well, allowing iPods to display business card information synced from a Mac.

### Second generation

A second generation iPod (2002)

The second generation iPod was introduced on July 17, 2002. Using the same body style as the first generation, the hold switch was redesigned, a cover was added to the FireWire port, and the mechanical wheel was replaced with a touch-sensitive wheel. The front plate also had rounded corners and edges. The second-generation class was available in 10 GB for US$399 and 20 GB for US$499. The first-generation 5 GB iPod was carried over, but its price was reduced to US$299.

Notably, Apple began selling Windows-compatible versions of the iPod starting with the second generation. These versions came with a 4-pin to 6-pin FireWire adapter and were bundled with Musicmatch Jukebox instead of iTunes.

In December 2002, Apple unveiled its first limited edition iPods, with either Madonna's, Tony Hawk's, or Beck's signature or No Doubt's band logo engraved on the back for an extra US$50.[5]

## Third generation

A 3rd Generation iPod (2003)

On April 29, 2003, Apple announced a completely redesigned third-generation iPod. Thinner than the previous models, the third generation models replaced the FireWire port with a new Dock Connector (which is still used today) and introduced the Touch Wheel, a completely non-mechanical interface with the four auxiliary buttons located in a row between the screen and the touch wheel. The front plate had rounded edges, and the rear casing was slightly rounded as well. A new wired remote connector was introduced. Whereas first and second generation iPods had an auxiliary ring around the headphone port for the remote, the third generation iPods had a 4-pin jack adjacent to the headphone port. A 10 GB model was sold for US$299, a 15 GB model for US$399, and a 30 GB model for US$499. All iPods were now compatible with Mac and Windows out of the box, simply requiring Windows users to reformat the iPod before use on a PC and both iTunes and Musicmatch were bundled with all iPods. The battery life was reduced to 8 hours, partially due to the use of a lithium-ion battery as opposed to a lithium polymer battery.

The 15 GB model was replaced by a 20 GB model and the 30 GB model was upgraded to 40 GB on September 8, 2003. The Windows-based Musicmatch software was made obsolete by and replaced by iTunes 4.1, the first version available for Microsoft Windows.

## Fourth generation

A 4th Generation iPod Photo (2004)

Announced on July 19, 2004, the fourth-generation iPod replaced the touch wheel from the third generation iPod with the Click Wheel from the iPod Mini, putting the four auxiliary buttons underneath a touch-sensitive scroll wheel. The casing was also slightly slimmer. Pricing was reduced and the lineup was simplified, as the 20 GB model was sold for US$299 and the 40 GB model for US$399. Notably, Apple began reducing pack-in accessories starting with the fourth generation. While a dock, carrying case, and wired remote were previously included with higher-end iPods, the higher-level 40GB iPod only came with a dock, earphones and an interchangeable proprietary cable capable of USB and FireWire interface. In addition to using the iPod Mini's Click Wheel, the fourth generation iPod used the more energy-efficient components of the Mini, allowing the fourth generation iPod to over 12 hours of battery life while using the same battery as its predecessor.

A special U2 edition was announced on October 26, 2004, to cross-market U2's *How to Dismantle an Atomic Bomb* album. The plastic front piece of the U2 edition iPod was black and the scroll wheel was red, to coincide with the color scheme of the U2 album. With 30 GB and the signatures of all four members of U2, the special edition iPod was priced at US$349 and also included a US$50 coupon for a US$149 collection of U2's entire back catalog. U2 iPods had special privileges, such as downloading songs for free.

A Special Harry Potter Edition was announced on September 7, 2005. It was released in conjunction with the Harry Potter Audiobooks in iTunes.[6] It had a Hogwarts logo engraved on the back, and all 6 Harry Potter Audiobooks which were available at the time preloaded.

**iPod Photo**

At the same time that the U2 iPod was announced, Apple also unveiled the iPod Photo.

Positioned as a premium version of the standard fourth-generation iPod, the iPod Photo featured a 220x176 pixel LCD capable of displaying up to 65,536 colors. The photo supported JPEG, BMP, GIF, TIFF, and PNG graphic file formats, and could be attached to a television or other external display for slideshows. The battery was rated for 15 hours for music playback and 5 hours for slideshows with music. The photo was available in a 40 GB version for US$499 and a 60 GB version for US$599.

On February 23, 2005, both 40 GB models (photo and regular) were replaced with a slimmer and lower-priced (US$349) 30 GB photo model leaving only a 20GB iPod regular left. The price for the 60 GB model was dropped to US$449, and accessory pack-ins were reduced, making the dock, FireWire cable, and television cable extra-cost options.

**iPod With Color Display**

On June 28, 2005, the iPod Photo was merged into the monochrome iPod line. The 30 GB model was dropped, and the 20 GB monochrome iPod received a color screen. The price for the 60 GB model was also dropped to US$399.

## Fifth generation

5th Generation iPod in white (2005).

The fifth generation iPod was introduced on October 12, 2005, shortly after the introduction of the iPod Nano. The fifth generation iPod featured a 2.5" 320x240 QVGA screen and a smaller Click Wheel. It was also known as the iPod Video and it was the first iPod to be able to play videos.

The iPod Video is the first iPod to be available in an alternative color scheme in a non-special edition form, as a black option was added alongside "Signature iPod White", and marked the second full redesign of the iPod's aesthetic with its re-arranged proportions, its return to a fully flat front plate, and its more rounded rear casing. The 4-pin remote port was removed as well, causing backwards compatibility issues with certain accessories. A 30 GB model was offered for US$299 and a 60 GB model was offered for US$399. The iPod Video was also offered in the U2 special edition for US$349 with 30 GB. The iPod Video was the last model to have a plastic face.

6G iPod Classic(left) & 5G iPod (right) iPod showing the updated view feature.

The iPod Video plays video in MP4 (up to 2.5 Mbit/s) and H.264 (up to 1.5 Mbit/s, baseline profile only) formats. Video such as TV shows, podcasts, music videos, and movies may be purchased from online stores such as the iTunes Store, or downloaded from Google Video and other sources, then imported to the iPod via iTunes software.

Videos or photo slideshows may be played from the fifth generation iPod on a television set, projector or monitor with the use of the Apple iPod AV cable or via a dock using an S-Video cable. It is also possible to do this using some camcorder cables with an RCA connection at one end and a three-banded eighth-inch (3.5 mm) A/V plug at the other, although the red and yellow plugs (normally the audio right and video signals respectively) must be swapped around in order to achieve the correct signal.

The iPod Video was updated on September 12, 2006. This update included a brighter screen, longer video playback time, newly designed earphones and a search feature. An iTunes installation CD was also no longer bundled, requiring users to download iTunes from Apple's website. The 60 GB model was replaced with an 80 GB model, and prices were cut by US$50 for both the 30 GB (US$249) and the 80 GB (US$349) models. Gapless playback and support for iPod games was enabled on all fifth generation iPods through a firmware update released at the same time.

### Sixth generation

6th Generation (2007).

During a special iPod-centric event on September 5, 2007, Steve Jobs introduced the sixth generation iPod and the suffix "classic". Featuring slightly thinner bodies, the sixth generation iPod also sported dramatically improved battery life, claiming up to 36 hours of music playback and 6 hours of video playback.[7] The iPod Classic has a 2.5" backlit display at a resolution of 320x240. The front plate of the iPod is now made of anodized aluminum instead of polycarbonate plastic, and "Signature iPod White" has been replaced by silver. This means that it is the first time that white is not a color option for an iPod in the iPod family. The sixth generation iPod also introduced a completely overhauled user interface, incorporating more graphics and Cover Flow. The iPod Classic was offered in an 80 GB model for MSRP US$249 and a 160 GB model for MSRP US$349. The U2 special edition has been dropped.

During the 'Let's Rock' Apple Event on September 9, 2008, the 80 GB and the thicker 160 GB model were discontinued in favor of a thin 120 GB version retailing for US$249. It introduced Genius and audio recording capabilities which are also available in other iPod models released at similar times; no firmware update provides either features to the first generation iPods.[8]

Prior to the 'It's Only Rock and Roll' event on September 9, 2009, the price of the 120 GB version was dropped to US$229. During the event Apple re-introduced the 160 GB version with the same slim profile as the 120 GB version, retailing at US$249. The 120 GB model has now been dropped and the only option available is the 160 GB model.

## Technical information

### User interface

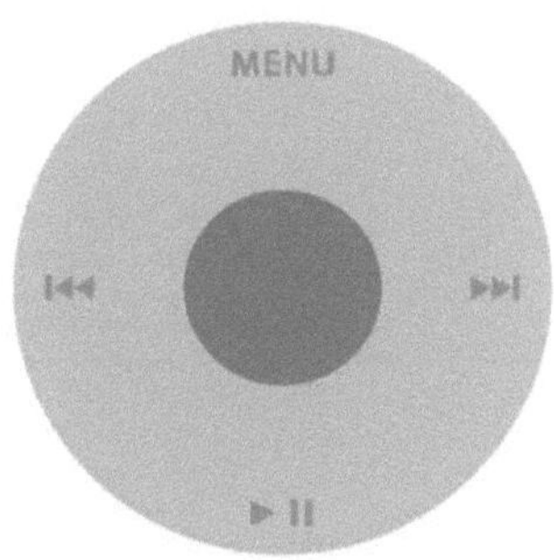

The iPod's signature Click Wheel

iPods with color displays use anti-aliased graphics and text, with sliding animations. All iPods have five buttons and the later generations (4th and above) have the buttons integrated into the click wheel — a design which gives an uncluttered, minimalist interface, though the circuitry contains multiple momentary button switches. The buttons are:

- *Menu*: to traverse backwards through the menus, toggle the backlight on older iPods, and jump to the main menu on newer iPods
- *Center*: to select a menu item
- *Play / Pause*: this doubles as an off switch when held
- *Skip Forward / Fast Forward*
- *Skip Backwards / Fast Reverse*

### Operating system and firmware

The iPod's operating system is stored on its dedicated storage medium. An additional NOR flash ROM chip (either 1 MB or 512 KB) contains a bootloader program that tells the device to load its OS from the storage medium. Each iPod also has 32 MB of RAM, although the 60 and 80 GB fifth generation, and the sixth generation models have 64 MB. A portion of the RAM is used to hold the iPod OS loaded from firmware, but the majority of it serves to cache songs from the storage medium. For example, an iPod could spin its hard disk up once and copy approximately 30 MB of upcoming songs into RAM, thus saving power by not requiring the drive to spin up for each song. Custom firmware has also been developed such as Rockbox (up to 6G - 6G requires emCORE) and iPodLinux (up to 5G) which offer open-source alternatives to the standard firmware and operating system.

### Additional features

In March 2002, Apple added limited PDA-like functionality: text files can be displayed, while contacts and schedules can be viewed and synchronized with the host computer.[9] Some built-in games are available, including *Brick* (a clone of Breakout), Parachute, *Solitaire*, and *Music Quiz*. A firmware update released in September 2006 brought some extra features to fifth generation iPods including adjustable screen brightness, gapless playback, and downloadable games. However, as of September 30, 2011, these games are no longer available on the iTunes Store.[10]

## Models

| Generation | Image | Capacity | Colours | Connection | Release date | Minimum OS to sync | Rated battery life (hours) |
|---|---|---|---|---|---|---|---|
| **first** |  | 5 GB | White | FireWire | October 23, 2001 | Mac: 9.2.1, 10.1 | *audio*: 10 |
| | | 10 GB | | | March 21, 2002 | | |
| | | First model, with mechanical scroll wheel. 10 GB model released later. Not compatible with Windows. | | | | | |
| **first (1st revision)** | | 5 GB | White | FireWire | July 17, 2002 | Mac: 9.2.2, 10.1.4<br>Win: 2000 | *audio*: 10 |
| | | | White (Limited Edition Madonna)<br>White (Limited Edition Tony Hawk)<br>White (Limited Edition No Doubt)<br>White (Limited Edition Beck) | | December 11, 2002 | | |
| | | Mechanical scroll wheel. Windows-compatible model available. Windows compatibility through Musicmatch. | | | | | |

| | | | | | | | |
|---|---|---|---|---|---|---|---|
| **second** |  | 10 GB | White | FireWire | July 17, 2002 | Mac: 9.2.2, 10.1.4<br>Win: 2000 | *audio*: 10 |
| | | | White (Limited Edition Madonna)<br>White (Limited Edition Tony Hawk)<br>White (Limited Edition No Doubt)<br>White (Limited Edition Beck) | | December 11, 2002 | | |
| | | 20 GB | White | | July 17, 2002 | | |
| | | | White (Limited Edition Madonna)<br>White (Limited Edition Tony Hawk)<br>White (Limited Edition No Doubt)<br>White (Limited Edition Beck) | | December 11, 2002 | | |
| | | Touch-sensitive wheel. FireWire port had a cover. Hold switch revised. Windows-compatible models available. Windows compatibility through Musicmatch. | | | | | |
| **third** |  | 10 GB | White | FireWire (USB for data transfer only — no charging) | April 28, 2003 | Mac: 10.1.5<br>Win: 2000 | *audio*: 8 |
| | | 15 GB | | | | | |
| | | 30 GB | | | | | |
| | | First complete redesign with all-touch interface, dock connector, and slimmer case. Musicmatch support dropped with later release of iTunes 4.1 for Windows. | | | | | |
| **third (1st revision)** | | 10 GB | White | FireWire (USB for data transfer only — no charging) | September 8, 2003 | Mac: 10.1.5<br>Win: 2000 | *audio*: 8 |
| | | 20 GB | | | | | |
| | | 40 GB | | | | | |
| **third (2nd revision)** | | 15 GB | White | FireWire (USB for data transfer only — no charging) | January 6, 2004 | Mac: 10.1.5<br>Win: 2000 | *audio*: 8 |
| | | 20 GB | | | | | |
| | | 40 GB | | | | | |
| **fourth** |  | 20 GB | White | FireWire or USB | July 19, 2004 | Mac: 10.1.5<br>Win: 2000 | *audio*: 12 |
| | | | Black/Red (Special Edition U2) | | October 26, 2004 | | |
| | | 40 GB | White | | July 19, 2004 | | |
| | | Adopted Click Wheel from iPod Mini; pack-in accessories reduced along with price drop. | | | | | |
| **fourth (photo)** |  | 40 GB | White | FireWire or USB | October 26, 2004 | Mac: 10.2.8<br>Win: 2000 | *audio*: 15<br>*slideshow*: 5 |
| | | 60 GB | | | | | |
| | | Premium spin-off of 4G iPod with color screen and picture viewing. | | | | | |
| **fourth (photo) (1st revision)** | | 30 GB | White | FireWire or USB | February 23, 2005 | Mac: 10.2.8<br>Win: 2000 | *audio*: 15<br>*slideshow*: 5 |
| | | 60 GB | | | | | |
| | | Pack-ins and price reduced. Images directly viewable via optional iPod Camera Connector. | | | | | |

| | | | | | | | |
|---|---|---|---|---|---|---|---|
| **fourth (with color display)** |  | 20 GB | White<br>Black/Red (Special Edition U2) | FireWire or USB | June 28, 2005 | Mac: 10.2.8<br>Win: 2000 | *audio*: 15<br>*slideshow*: 5 |
| | | | White (Special Edition Harry Potter) | | September 7, 2005 | | |
| | | 60 GB | White | | June 28, 2005 | | |
| | | "iPod with color display"; essentially, the iPod Photo model reintegrated with the main iPod lineup. | | | | | |
| **fifth** | 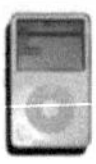 | 30 GB | White<br>Black<br>White (Special Edition Harry Potter)<br>Black (Special Edition Harry Potter) | USB (FireWire for charging only) | October 12, 2005 | Mac: 10.3.9<br>Win: 2000 | *audio*: 14<br>*slideshow*: 3<br>*video*: 2 |
| | | | Black/Red (Special Edition U2) | | June 6, 2006 | | |
| | | 60 GB | White<br>Black | | October 12, 2005 | | *audio*: 20<br>*slideshow*: 4<br>*video*: 3 |
| | | Second full redesign with a slimmer case, and larger screen with video playback. Offered in black or white. | | | | | |
| **fifth (1st revision)** | | 30 GB | White<br>Black<br>Black/Red (Special Edition U2) | USB (FireWire for charging only) | September 12, 2006 | Mac: 10.3.9<br>Win: 2000 | *audio*: 14<br>*slideshow*: 4<br>*video*: 3.5 |
| | | 80 GB | White<br>Black | | | | *audio*: 20<br>*slideshow*: 6<br>*video*: 6.5 |
| | | Battery life improved for slideshow and video playbacks as well as a very slight change in software giving the user the "search" ability. | | | | | |
| **sixth (classic)** |  | 80 GB | Silver<br>Black (black front plate and black wheel) | USB (FireWire for charging only) | September 5, 2007 | Mac: 10.4.8<br>Win: XP SP2 | *audio*: 30<br>*video*: 5 |
| | | 160 GB | | | | | *audio*: 40<br>*video*: 7 |
| | | Introduced the "classic" suffix. New interface and anodized aluminum front plate. Silver replaces white. | | | | | |
| **sixth (classic) (1st revision)** | | 120 GB | Silver<br>Black (gray front plate and black wheel) | USB (FireWire for charging only) | September 9, 2008 | Mac: 10.4.11<br>Win: XP SP3 | *audio*: 36<br>*video*: 6 |
| | | Genius feature added. 160 GB model dropped and 80 GB model upgraded to 120 GB. Can use mic and remote controller through 3.5mm audio jack. Front plate color changed (Black to Gray). | | | | | |
| **sixth (classic) (2nd revision)** | | 160 GB | Silver<br>Black (smoke front plate and black wheel) | USB (FireWire for charging only) | September 9, 2009 | Mac: 10.4.11<br>Win: XP SP3 | *audio*: 36<br>*video*: 6 |
| | | Capacity increased to 160 GB using single-platter drive. Added Genius Mixes (after Software Update). | | | | | |

# Timeline of full-size iPod models

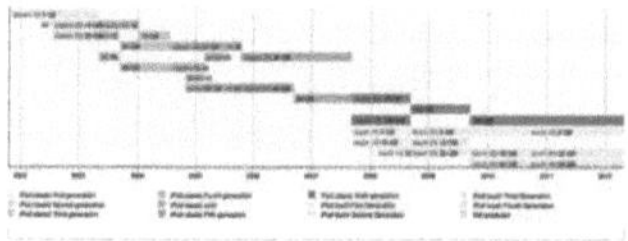

*Sources: Apple press release library,*[11] *Mactracker Apple Inc. model database*[12]

# References

[1] "Ebay Listing for "Seventh Generation" iPod Classic" (http://www.ebay.com/ctg/Apple-iPod-classic-7th-Generation-Black-160-GB-Latest-Model-/85890740). Ebay.com. . Retrieved 2011-10-23.

[2] "Broken? Sell It on eBay: An iPod Report" (http://www.auctionbytes.com/cab/abu/y206/m05/abu0167/s04). Auctionbytes.com. . Retrieved 2011-10-23.

[3] Apple Computer, Inc. (1999-03-26). "Apple – QuickTime – September 2007 Keynote Address" (http://events.apple.com.edgesuite.net/s83522y/event/index.html?internal=g4h5jl83a). Events.apple.com.edgesuite.net. . Retrieved 2011-10-23.

[4] "Apple Support Article – Identifying iPod Models" (http://support.apple.com/kb/HT1353). Support.apple.com. 2011-10-13. . Retrieved 2011-10-23.

[5] Dalrymple, Jim. Limited Edition Madonna, Tony Hawk, Beck iPods (http://www.macworld.com/news/2002/12/10/ipod/index.php). *Macworld*, September 10, 2002. Retrieved on January 7, 2007.

[6] Harry Potter Digital Audiobooks Debut Exclusively on iTunes Music Store (http://www.apple.com/pr/library/2005/sep/07potter.html) Apple.com

[7] "Apple – iPod classic – Read the iPod classic technical specifications" (http://www.apple.com/ipodclassic/specs.html). Apple.com. . Retrieved 2011-10-23.

[8] "Apple iPod classic review" (http://reviews.cnet.com/mp3-players/apple-ipod-classic-second/4505-6490_7-33247504.html). *CNet.com*. .

[9] "Apple Introduces 10 GB iPod—2,000 Songs in Your Pocket" (http://www.apple.com/pr/library/2002/mar/20ipod.html). Apple Inc. March 20, 2002. . Retrieved February 18, 2007.

[10] "Apple removes click-wheel games from iTunes as iPod classic lives its last days" (http://www.appleinsider.com/articles/11/09/30/apple_removes_ipod_classic_click_wheel_games_from_itunes_store.html). AppleInsider. 2011-09-30. . Retrieved 2011-10-23.

[11] Apple Inc., Apple press release library (http://www.apple.com/pr/library/), Retrieved September 19, 2007.

[12] Mactracker ( mactracker.ca (http://www.mactracker.ca/)), Apple Inc. model database, version as of 26 July 2007.

# External links

- Official website (http://http://www.apple.com/ipodclassic/)
- http://support.apple.com/specs/#ipodclassic

# iPod Touch

Black 4th generation iPod Touch

| | |
|---|---|
| **Developer** | Apple Inc. |
| **Manufacturer** | Foxconn |
| **Release date** | • September 8, 2010 (current generation)<br>• September 5, 2007 (1st generation) |
| **Units sold** | 60 million |
| **Operating system** | 1G: iOS 3.1.3<br>2G: iOS 4.2.1<br>3G: iOS 5.1<br>4G: iOS 5.1 |
| **Power** | 3.7 V rechargeable lithium-ion battery, 3.44 Wh, 930 mAh;<br>audio: 40 hours; video: 7 hours[1] [2] |
| **CPU** | ARM Cortex-A8 Apple A4 1 GHz (underclocked to 800 MHz) |
| **Storage capacity** | 8, 32 or 64 GB flash memory |
| **Memory** | 256 MB DRAM[3] |
| **Display** | 3.5 in (**unknown operator: u'strong'** mm), 3:2 aspect ratio, 24-bit color depth, aluminosilicate glass-covered LED-backlit LCD, 960×640 px at 326 PPI |
| **Input** | • Multi-touch touchscreen display<br>• ambient light sensors<br>• volume buttons<br>• microphone<br>• voice control<br>• 3-axis accelerometer, gyroscope |
| **Camera** | • Rear: Video recording, 1280×720 up to 30 frames/s with audio, 960×720 still camera with 5× digital zoom, 0.7 MP[4] [5]<br>• Front: Video recording, VGA up to 30 frames/s with audio, VGA-quality still camera, 0.3 MP |
| **Connectivity** | • Wi-Fi 802.11b/g/n (802.11n 2.4 GHz only)<br>• USB 2.0 dock connector<br>• Bluetooth 2.1[6] |
| **Online services** | App Store, iTunes Store, iBookstore, iCloud |

| | |
|---|---|
| **Dimensions** | 110 mm (**unknown operator: u'strong'** in) H<br>58 mm (**unknown operator: u'strong'** in) W<br>7.1 mm (**unknown operator: u'strong'** in) D |
| **Weight** | 101 g (**unknown operator: u'strong'** oz) |
| **Related articles** | List of iOS devices |
| **Website** | www.apple.com/ipodtouch [7] |

The **iPod Touch** (stylized and marketed as lowercase **iPod touch**) is a portable media player, personal digital assistant, handheld game console, and Wi-Fi mobile device designed and marketed by Apple Inc. The iPod Touch adds the multi-touch graphical user interface to the iPod line. It is the first iPod with wireless access to the iTunes Store, and also has access to Apple's App Store, enabling content to be purchased and downloaded directly on the device. As of March 2011, Apple has sold over 60 million iPod Touch units.[8] [9]

## Software

The iPod Touch runs iOS. The first major update after the initial release was iPhone OS 2.0 This update introduced the App Store, which allowed third-party applications for the first time. iPhone OS 2.0 debuted July 11, 2008. iPhone users received the update for free, while iPod Touch users had to pay for the update. The second major update to the operating system, iPhone OS 3.0, was released June 17, 2009. iPhone OS 3.0 added features such as cut, copy, and paste; data tethering; and push notification support. As with the prior major release, iPhone users received the update for free, while iPod Touch users had to pay for the update. iOS 4.0 was made available to the public on June 21, 2010. It was the first major iOS release to drop support for some devices; the first generation iPod Touch and original iPhone are not supported in iOS 4.0. The iPhone 3G and second generation iPod Touch had limited functionality under iOS 4.0, while the iPhone 4, iPhone 3GS, third generation iPod Touch, and fourth generation iPod Touch had full functionality under iOS 4.0. The major features introduced in iOS 4.0 included iBooks, FaceTime, and multitasking. iOS 5.0 was previewed to the public on June 6, 2011, and was released on October 12, 2011.[10]

Apple has received criticism for charging iPod Touch owners for major iOS updates that iPhone owners receive for free, as well as excluding certain features from the iPod Touch software that are included in the iPhone.[11] [12] Apple has said that they can add features for free to the iPhone because the revenue from it is accounted for on a subscription basis under accounting rules, rather than as a one time payment.[13] At WWDC in June 2010, Steve Jobs announced that Apple had "found a way" to make subsequent OS upgrades available free to iPod Touch owners. However, iOS 4.0 and onwards isn't available for the first generation iPod Touch, and some features are only available for the 32 and 64 GB third generation iPod Touch released in September 2009 and the fourth generation iPod Touch released in 2010.

In June 2011, iOS 5, the fifth major release of iOS software, was announced at Apple's WWDC 2011. There will be added features in the new update, including notifications, messaging and reminders. On October 4, 2011 at Apple's "Let's Talk iPhone" event iOS 5 and the iPhone 4S were announced to the public with many new features such as Siri, iMessage and wireless upgrades; this software upgrade was released on October 12, 2011. No new iPod Touch specific changes were announced yet, other than a price drop and the option to choose between black and white.

## Comparison to the iPhone

The iPod Touch and the iPhone share essentially the same hardware and run the same iOS operating system. The iPod Touch lacks some of the iPhone's features and associated apps, such as access to cellular networks, GPS navigation and the built-in compass. Older models also lacked speakers and cameras. Although the SMS and Phone apps are included on the iPod Touch software, they are disabled and therefore not visible. Also, the sleep/wake button was on the opposite side, up until the release of the iPod Touch 4th generation. Since it doesn't need GPS and cellular components, the iPod Touch is slimmer and lighter than the iPhone. Steve Jobs once referred to the iPod Touch as "training wheels for the iPhone".[14]

Another major difference is the quality of the back camera compared to the iPhone. While the iPod Touch, like the iPhone, allows for HD video recording, the iPhone camera still delivers higher quality photos to that of the iPod's.[15] Also, the iPod Touch camera lacks a flash for low light photography.

## Requirements

- iTunes 10 or later (iTunes 10.5 for iOS 5.0+)
- Mac OS 10.5 or later (10.5.8 for iOS 5.0+)
- Windows XP Home or Professional with Service Pack 3 or later (Vista to run iCloud)

### Setup and synchronization

For iPod Touch units bought before October 12, 2011, users must own a Mac or PC computer to be able to use the iPod. Users then must install iTunes and connect the iPod through a USB port. The iPod will then be set up in iTunes. New iPods bought after October 12, 2011 have iOS 5.0 preloaded, and allow activation wirelessly, without the need of a PC or Mac.[16]

Earlier iPod Touch units must be plugged into a computer to be synced. This will charge the iPod Touch and sync the music library, videos, pictures and back up data. iOS 5 enables the user to do all this from the device and send it to the iCloud service. It is not manageable by Mac OS X 10.4 Tiger.[17]

### Battery charging

Starting with the second generation, iPod Touch can only be charged from the 5 V pin of the dock connector while most prior iPod models (including the original iPod Touch) could also be charged from the 12 V pin for FireWire power.[18] This change dropped support for charging in vehicles equipped with a FireWire-based iPod connection. Most aftermarket manufacturers of such equipment offer cables and/or adapters which convert the vehicle's 12 V to 5 V.

### Purchasing content

To purchase content on the iPod Touch, the user must create an Apple ID or have an existing account. With this account you may download music and videos from the iTunes Store, apps from the App Store, or books from the iBookstore. An Apple ID account created without a credit card can be used to get free content.. If you have a problem with your iPod, you better expect to pay up in order to get some help. Apple gives you one free call to their support line, but it has to be used within the first 90 days of your purchase. Calling for support 90 days after your purchase will cost $49.99.[19]

### Third-party applications

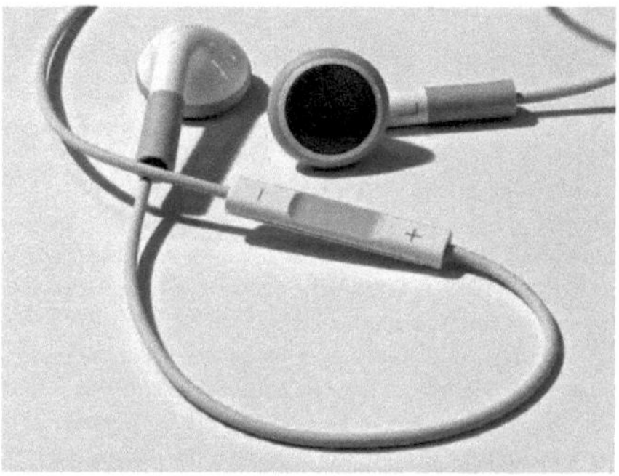

Earphones with an external microphone and controller are compatible with the 2nd generation iPod Touch, and are shipped with the 3rd generation iPod Touch devices (excluding the 8 GB model).

The only official way to obtain third-party applications for the iPod Touch is Apple's App Store, which is a branch of iTunes Store. The App Store application, available in all versions of iOS from 2.0 onwards, allows users to browse and download applications from a single online repository (hosted by Apple) with the iTunes Store. To develop such software, a software development kit (SDK) was officially announced on March 6, 2008, at an Apple Town Hall meeting.[20] The iOS SDK allows making applications for the iPhone and iPod Touch after paying a fee to join the development team. The developer can then set the price for the applications they develop and will receive 70% of money earned. Apple retains 30% of the sale price with no minimum fixed fee.

## Hacks

A month after the first iPod Touch was released in September 2007, hackers released JailbreakMe 1.0 (also called "AppSnapp") to jailbreak iPhone OS 1.1.1 on the iPhone and iPod Touch,[21] [22] enabling users to install third-party programs on their devices (otherwise unavailable since the App Store was not part of iOS yet). These third-party programs could use additional functionality not supported by Apple, such as enabling multitasking, applying themes to the home screen, or enabling a battery percentage indicator. As of December 2011, each version of iOS on iPod Touches can be jailbroken, from 1.1.1 to 5.0.1 (with a tethered jailbreak for versions 4.3.4 through 5.0), using various tools including PwnageTool and redsn0w (see iOS jailbreaking for details). Apple's warranty statement implies that an iPod Touch after jailbreaking or other modification made by unofficial means is not covered by Apple's warranty. However, such disclaimers are generally considered invalid, due to consumer protection laws. Furthermore, the jailbreaking process is easily undone by performing a restore through iTunes).[23] [24]

## Models

| Model | Picture | Capacity | RAM | Connection | Original release date | Minimum OS to sync | Rated battery life (hours) |
|---|---|---|---|---|---|---|---|
| 1st generation |  | 8 GB<br>16 GB<br>32 GB[25] | 128 MB[26] | USB via dock connector (FireWire for charging only)[18] | September 5, 2007 | Mac: 10.4 or later<br>Win: XP, Vista, 7 | *audio*: 22<br>*video*: 5 |
| | The first generation iPod Touch was launched on September 5, 2007, at an event called *The Beat Goes On.*[27] First iPod with Wi-Fi and a multi-touch interface. Features Safari web browser and wireless access to the iTunes Store and YouTube. Later added 16 and 32 GB versions, 32 GB in Feb 2008. iPhone OS 2.0 and App Store access require an upgrade fee. iOS 4.0 and later are not supported because they require newer hardware. | | | | | | |

| | | | | | | | |
|---|---|---|---|---|---|---|---|
| **2nd generation** |  | 8 GB<br>16 GB<br>32 GB[25] | 128 MB[26] | USB via dock connector | September 9, 2008 | Mac: 10.4 or later<br>Win: XP, Vista, 7 | *audio*: 36<br>*video*: 6 |
| | The second generation iPod Touch featured a new tapered chrome back with Nike+ functionality, volume buttons, and built-in speaker added, It was unveiled on September 9, 2008, at the "Let's Rock" keynote presentation. iPhone OS 2.0 and App Store access standard. Bluetooth support added but not made active until iPhone OS 3.0, which requires an upgrade fee. Support for external microphone and remote added. The 8 GB model was re-introduced alongside the 3rd generation iPod Touch in September 2009, and included the newer iPhone OS 3.1. iOS 4 was released for free for second generation iPod Touch models June 21, 2010, but included a limited selection of features. Future iOS software updates no longer supported (as of iOS 4.2.1). The 2nd generation also comes with the chrome frame seen on the iPhone 3G, making the two almost identical when viewed from the front. However, there are some notable differences, as there is no speaker above the screen, no silent/ringer switch, the back is metal, and the sleep/wake button is on the other side. The 2nd generation also supports audio input when a headphone or earphone with microphone abilities is plugged into the audio output (headphone) jack. The iPod Touch 2.0 Software Update supports WPA2 Enterprise with 802.1X authentication.[28] The iPhone OS 3.0 update unlocked Bluetooth ability on the 2nd generation iPod Touch, as the included Wi-Fi chip (Broadcom BCM4325) has Bluetooth support.[29] [30] The 2nd generation iPod Touch is said to have a yellower cast/tint to the display, as compared to the iPhone or the original iPod Touch.[31] The applications processor inside the 2nd generation iPod Touch runs slightly faster than the processor inside the iPhone 3G, but slower than the iPhone 3GS.[32] [33] The first generation iPod Touch works with all "Made for iPod" peripherals, but certain changes that Apple made to the 2nd generation iPod Touch prevent some existing peripherals from recharging the updated player.[34] The Google Street View feature added on iPhone firmware version 2.2 is absent from the same version of firmware released on the iPod Touch,[35] but is found in the 3.0 update. The 2nd and 3rd generation iPod Touch can be updated to iOS 4.x for free, but cannot take advantage of features using the iPhone's camera or GPS receiver. The 2nd generation iPod Touch is further unable to take advantage of iOS 4's multitasking and home screen background customization features, unless the firmware is modified (through jailbreaking), and cannot be upgraded beyond iOS version 4.2.1. Later updates of iOS beyond 4.2.1 are no longer supported due to the outdated hardware. The iPod Touch (2nd generation) is model A1288. | | | | | | |
| **3rd generation** |  | 32 GB<br>64 GB[25] | 256 MB | USB via dock connector | September 9, 2009 | Mac: 10.4 or later<br>Win: XP, Vista, 7 | *audio*: 30<br>*video*: 6 |
| | The 32 GB and 64 GB models were updated to include the upgraded RAM, CPU and GPU internals from the iPhone 3GS; they include voice control support, VoiceOver [36], included iPhone OS version 3.x, and bundled remote earphones with microphone. The 3rd generation iPod Touch with iPhone OS 3.1 was announced and subsequently released on September 9, 2009.[37] Apple's published technical specifications for the 3rd generation iPod Touch only include references to the 32 GB and 64 GB models.[38] Following the introduction of the 3rd generation iPod Touch, Apple also continued offering the 2nd generation 8 GB version, upgraded to the same iPhone OS version as the new 3rd generation devices. The iPod Touch (3rd generation) is model A1318. Apple as well sold many refurbished iPod Touch 2nd generation models with the 3rd generation software effectively making them a 3rd generation unit. These can be identified by software version 4.3.5. | | | | | | |
| **4th generation** |  | 8 GB<br>32 GB<br>64 GB[25] | 256 MB[39] | USB via dock connector | September 1, 2010 | Mac: 10.5 or laterWin: XP, Vista, 7 | *audio*: 40<br>*video*: 7 |
| | Apple unveiled the 4th generation iPod Touch on September 1, 2010. It features a front-facing camera for FaceTime, the iPhone 4's Retina display without an IPS screen,[40] support for recording 720p video and 960 x 720 still photos via a back camera,[5] Apple's A4 chip, a built-in microphone, and a 3-axis gyroscope. The display is glossier than in previous generations. On October 12, 2011, the iPod Touch 4th-generation began shipping with iOS 5.0 pre-installed for PC-free activation. Also, an additional color option, white, was made available. | | | | | | |

## Specifications

### Legend

| Discontinued | Current |
|---|---|

| Model | 1st generation | 2nd generation | 3rd generation | 4th generation |
|---|---|---|---|---|
| **Pre-installed OS** | iPhone OS 1.1 | iPhone OS 2.1.1<br>iOS 3.1.1 (updated 2G 8 GB MC model) | iPhone OS 3.1.1 | iOS 4.1 (black)<br>iOS 5.0 (black & white) |
| **Highest Supported OS** | iPhone OS 3.1.3 | iOS 4.2.1 | iOS 5.1 | |
| **Display** | 3.5 in (**unknown operator: u'strong'** mm); 2:3 aspect ratio, 18-bit color depth, glossy glass-covered LED-backlit LCD screen, 480×320 px (HVGA) at 163 PPI | | | 3.5 in (**unknown operator: u'strong'** mm); 2:3 aspect ratio; emulated 24-bit color depth(18-bit + dithering), glossy glass-covered LED-backlit LCD, 960x640 px at 326 PPI |
| **System on chip** | Samsung S5L8900 | Samsung S5L8720 | Samsung S5L8920 | Apple A4[41] (Samsung S5L8930) |
| **CPU core** | 620 MHz (underclocked to 412 MHz, originally 400 MHz)<br>ARM 1176JZ(F)-S[42] | 620 MHz (underclocked to 533 MHz)<br>ARM 1176JZ(F)-S[42] | 833 MHz (underclocked to 600 MHz)<br>ARM Cortex-A8 core[26] | 1 GHz (underclocked to 800 MHz)<br>ARM Cortex-A8 |
| **GPU core** | PowerVR MBX Lite GPU[43] [44] | | PowerVR SGX535 GPU[45] | |
| **Storage** | 8, 16 or 32 GB | | 32 or 64 GB[46] | 8, 32 or 64 GB |
| **Memory** | 128 MB DRAM[47] | | 256 MB DRAM[3] [41] [48] | |
| **Connectivity** | Wi-Fi (802.11b/g),<br>USB 2.0 dock connector | In addition to prior:<br>Bluetooth 2.1 + EDR (requires iPhone OS 3.0),<br>Built-in speaker, hardware volume controls, Nike+ | In addition to prior:<br>Voice control,<br>Includes earphones with remote and mic | In addition to prior:<br>802.11n (2.4 GHz only)<br>3-axis gyroscope<br>Microphone |
| **Cameras** | N/A | | | Back (main): CMOS image sensor with video (720p HD at 30 frames/s);<br>Front: VGA-quality photos and video at up to 30 frames/s. |
| **Audio processor** | Wolfson Microelectronics WM8758BG[49] | Cirrus Logic CS4398 | Cirrus Logic CS4398[50] | Cirrus Logic |
| **Materials** | Glass display, stainless steel back and aluminum bezel; plastic for Wi-Fi antenna | Glass display, contoured stainless steel back and bezel; plastic for Wi-Fi and Bluetooth antenna | | Glass display, stainless steel back and bezel |
| **Color** | Black only | | | Black or white |
| **Power** | Built-in rechargeable lithium-ion polymer battery[51] [52] [53] | | | |
| | 3.7 V 2.15 W·h (580 mA·h) | 3.7 V 2.73 W·h (739 mA·h)[52] [54] | 3.7 V 2.92 W·h (789 mA·h)[53] | 3.7 V 3.44 W·h (930 mA·h) |
| **Rated battery life (hours)** | *audio*: 22<br>*video*: 5 | *audio*: 36<br>*video*: 6 | *audio*: 30<br>*video*: 6 | *audio*: 40<br>*video*: 7 |

<table>
<tr><td>Dimensions</td><td>110 mm (unknown operator: u'strong' in) H<br>61.8 mm (unknown operator: u'strong' in) W<br>8 mm (unknown operator: u'strong' in) D</td><td colspan="2">110 mm (unknown operator: u'strong' in) H<br>61.8 mm (unknown operator: u'strong' in) W<br>8.5 mm (unknown operator: u'strong' in) D</td><td>111 mm (unknown operator: u'strong' in) H<br>58.9 mm (unknown operator: u'strong' in) W<br>7.2 mm (unknown operator: u'strong' in) D</td></tr>
<tr><td>Weight</td><td>120 g (unknown operator: u'strong' oz)</td><td colspan="2">115 g (unknown operator: u'strong' oz)</td><td>101 g (unknown operator: u'strong' oz)</td></tr>
<tr><td>Released</td><td>8, 16 GB: September 14, 2007<br>32 GB: February 5, 2008</td><td>September 9, 2008</td><td>September 9, 2009</td><td>Black: September 8, 2010<br>White: October 12, 2011</td></tr>
<tr><td>Discontinued status</td><td>September 9, 2008</td><td>16, 32 GB:<br>September 9, 2009<br>8 GB:<br>September 1, 2010</td><td>September 12, 2010</td><td>Current</td></tr>
</table>

## See also

- Comparison of iPod managers
- Comparison of portable media players
- List of portable media players with Wi-Fi connectivity
- List of iOS devices

## References


[1] FCC Office of Engineering and Technology. "OET List Exhibits Report" (https://fjallfoss.fcc.gov/oetcf/eas/reports/ViewExhibitReport.cfm?mode=Exhibits&RequestTimeout=500&calledFromFrame=N&application_id=338019&fcc_id='BCG-E2407). US Federal Communications Commission. . Retrieved July 5, 2011.

[2] Horwitz, Jeremy (September 1, 2010). "iPod touch 4G's Battery Is Bigger, But Is It Better?" (http://www.ilounge.com/index.php/backstage/comments/ipod-touch-4gs-battery-is-bigger-but-is-it-better/). iLounge. . Retrieved February 19, 2011.

[3] "Apple's iPod Touch (2010) Review, Not a Poor Man's iPhone 4" (http://www.anandtech.com/show/3903/apples-ipod-touch-2010-review-not-a-poor-mans-iphone-4). AnandTech. September 8, 2010. . Retrieved September 8, 2010.

[4] iPod Touch 4th Generation (September 8, 2010). "iPod Touch 4th Generation Teardown" (http://www.ifixit.com/Teardown/iPod-Touch-4th-Generation-Teardown/3562/1#s17091). iFixit. . Retrieved July 5, 2011.

[5] "iPod touch – Technical specifications for iPod touch" (http://www.apple.com/ipodtouch/specs.html). Apple. . Retrieved July 5, 2011.

[6] "iPod touch gets Bluetooth after OS upgrade" (http://news.cnet.com/ipod-touch-gets-bluetooth-after-os-upgrade/). *CNET*. CBS Interactive. . Retrieved January 22, 2010.

[7] http://www.apple.com/ipodtouch

[8] "Apple's Samsung lawsuit reveals over 60 million iPod touch sold" (http://www.appleinsider.com/articles/11/04/19/apples_samsung_lawsuit_notes_over_60_million_ipod_touch_sold.html). AppleInsider. April 19, 2011. . Retrieved July 5, 2011.

[9] "Apple sues Samsung: a complete lawsuit analysis" (http://thisismynext.com/2011/04/19/apple-sues-samsung-analysis/). This is my next.... . Retrieved July 5, 2011.

[10] "Apple announce iOS 5 and iPhone release date" (http://events.apple.com.edgesuite.net/11piuhbvdlbkvoih10/event/index.html). Apple. October 4, 2011. . Retrieved October 4, 2011.

[11] "iPod touch users: if you bought after Jan 1 the new apps are free – maybe" (http://www.engadget.com/2008/01/17/ipod-touch-users-if-you-bought-after-jan-1-the-new-apps-are-fre/). Engadget.com. . Retrieved March 3, 2010.

[12] Arya, Aayush (January 24, 2008). "Early Adopter Tax Resurfaces with the iPod touch January Software Upgrade" (http://www.applematters.com/article/early-adopter-tax-resurfaces-with-the-ipod-touch-january-software-upgrade/). AppleMatters. . Retrieved March 3, 2010.

[13] Dalrymple, Jim. "Accounting rules behind iPod touch update charge" (http://www.macworld.com/article/131991/2008/02/ipodtouch.html). Macworld. . Retrieved March 3, 2010.

[14] Fildes, Nic (September 19, 2007). "iPhone finally arrives but it's neither cheap nor G3." (http://www.independent.co.uk/news/business/news/iphone-finally-arrives-but-its-neither-cheap-nor-g3-402835.html). *The Independent* (UK). . Retrieved October 24, 2008.

[15] Foresman, Chris. "iPod Touch camera review" (http://arstechnica.com/apple/reviews/2010/09/ipod-touch-4-vs-iphone-4-camera-fight-no-contest.ars). *iPod touch camera review*. .

[16] "iOS 5 - See new features included in iOS 5" (http://www.apple.com/ios/features.html#pcfree). Apple. . Retrieved 2011-11-29.

[17] View: Everyone Only Notes (September 2, 2010). "iPod touch 4G on Mac OS X 10.4?: Apple Support Communities" (https://discussions.apple.com/thread/2565323?threadID=2565323). Apple Support Communities. . Retrieved July 5, 2011.

[18] "iPhone and iPod touch: Charging the battery" (http://support.apple.com/kb/HT1476). Apple. October 15, 2008. . Retrieved January 12, 2010.

[19] http://www.worldstart.com/interesting-ipod-facts/

[20] "Live from Apple's iPhone SDK press conference" (http://www.engadget.com/2008/03/06/live-from-apples-iphone-press-conference/). Engadget. . Retrieved March 3, 2010.

[21] Wilson, Ben (2007-10-10). "Official iPhone 1.1.1 jailbreak released with easy-to-follow instructions; does not require TIFF exploit" (http://reviews.cnet.com/8301-19512_7-10115200-233.html). *CNET*. CBS Interactive. . Retrieved 2009-11-10.

[22] Gregg Keizer (October 29, 2007). "Hacker Software Can Install Unauthorized Software on iPhones" (http://www.pcworld.com/article/139061/hacker_software_can_install_unauthorized_software_on_iphones.html). PCWorld. . Retrieved August 25, 2011.

[23] http://wiki.eeeuser.com/warranty

[24] iPod and iSight Warranty (PDF) (http://images.apple.com/legal/warranty/docs/ipodisight.pdf), Apple, page 1. Retrieved on December 24, 2008.

[25] "Identifying iPod models" (http://support.apple.com/kb/ht1353). Apple Inc.. . Retrieved February 13, 2011.

[26] "The iPhone 3GS Hardware Exposed & Analyzed" (http://www.anandtech.com/gadgets/showdoc.aspx?i=3579&p=2). AnandTech. . Retrieved June 7, 2010.

[27] Steve Jobs live – Apple's "The heat is on" special event (http://www.engadget.com/2007/09/05/steve-jobs-live-apples-the-heat-is-on-special-event/). Retrieved September 5, 2007.

[28] "Apple Announces iPhone 2.0 Software Beta" (http://www.apple.com/pr/library/2008/03/06iphone.html). *Apple Press Info*. Apple Inc. March 6, 2008. . Retrieved May 15, 2008.

[29] "Broadcom page on wireless chip included in the 2nd generation iPod Touch" (http://www.broadcom.com/products/Bluetooth/Bluetooth-RF-Silicon-and-Software-Solutions/BCM4325). Broadcom. .

[30] "iPod Touch gets Bluetooth after OS upgrade" (http://news.cnet.com/8301-17938_105-10198524-1.html?tag=TOCmoreStories.0). *CNET*. CBS Interactive. .

[31] Breen, Christopher. "First Look: iPod touch" (http://www.macworld.com/article/135484/2008/09/ipodtouchfirstlook.html?lsrc=rss_main). Macworld. . Retrieved March 3, 2010.

[32] Krazit, Tom (November 24, 2008). "New iPod Touch faster than iPhone 3G" (http://news.cnet.com/8301-13579_3-10106891-37.html). *CNET*. CBS Interactive. .

[33] "2nd Generation iPod Touch Faster than iPhone" (http://toucharcade.com/2008/11/23/2nd-generation-ipod-touch-faster-than-iphone/). Touch Arcade. . Retrieved March 3, 2010.

[34] iPod touch 2G Won't Support Some iPod Peripherals (http://www.mp3newswire.net/stories/8002/ipod-touch-peripheral.htm) – MP3 Newswire

[35] "iPod Touch Denied Street View, Other Google Maps Upgrades in Firmware 2.2 – iPhone" (http://gizmodo.com/5095448/ipod-touch-denied-street-view-other-google-maps-upgrades-in-firmware-22). Gizmodo. November 21, 2008. . Retrieved March 3, 2010.

[36] http://www.apple.com/accessibility/itunes/ipodtouch.html

[37] "Apple Introduces New iPod touch Lineup" (http://www.apple.com/pr/library/2009/09/09touch.html). Apple. September 9, 2008. . Retrieved March 2, 2010.

[38] "iPod touch (3rd generation) – Technical Specifications" (http://support.apple.com/kb/SP570). Apple. April 8, 2010. . Retrieved March 15, 2011.

[39] "iPod Touch 4th Generation Teardown" (http://www.ifixit.com/Teardown/iPod-Touch-4th-Generation-Teardown/3562/3). iFixit. September 8, 2010. p. 3. . Retrieved February 19, 2011.

[40] Topolsky, Joshua (September 7, 2010). "iPod touch review (2010)" (http://www.engadget.com/2010/09/07/ipod-touch-review-2010/). Engadget. . Retrieved February 19, 2011.

[41] "iPod Touch 4th Generation Teardown – Page 3" (http://www.ifixit.com/Teardown/iPod-Touch-4th-Generation-Teardown/3562/3). iFixit. . Retrieved September 15, 2010.

[42] "That iPod Touch runs at 533 MHz" (http://www.pcworld.com/article/154518/that_ipod_touch_runs_at_533_mhz.html). *PC World*. 2008-11-25. . Retrieved 2011-07-31.

[43] "The iPhone 3GS Hardware Exposed & Analyzed" (http://www.anandtech.com/gadgets/showdoc.aspx?i=3579&p=3&cp=6). AnandTech. . Retrieved June 7, 2010.

[44] "Intel ups stake in iPhone GPU maker" (http://www.electronista.com/articles/08/12/23/intel.stake.in.imagination/). Electronista.com. December 23, 2008. . Retrieved March 3, 2010.

[45] "Apple iPhone 3G S – OpenGL ES performance and system information" (http://glbenchmark.com/phonedetails.jsp?benchmark=glpro11&D=Apple iPhone 3G S&testgroup=gl). Glbenchmark.com. . Retrieved June 7, 2010.

[46] "iPod touch (3rd generation) - Technical Specifications" (http://support.apple.com/kb/SP570). Support.apple.com. 2010-04-08. . Retrieved 2011-11-29.

[47] "The iPhone 3GS Hardware Exposed & Analyzed" (http://www.anandtech.com/gadgets/showdoc.aspx?i=3579&p=1). AnandTech. . Retrieved June 7, 2010.

[48] "iPhone 3GS gets PowerVR SGX GPU core, double the RAM" (http://www.electronista.com/articles/09/06/10/ipgone.3gs.gets.gpu.ram/). Electronista. June 10, 2009. . Retrieved June 7, 2010.
[49] "iPod Touch 1st Generation Teardown – Page 3" (http://www.ifixit.com/Teardown/iPod-Touch-1st-Generation-Teardown/596/3). iFixit. . Retrieved June 25, 2010.
[50] "iPod Touch 3rd Generation Teardown" (http://www.ifixit.com/Teardown/iPod-touch-3rd-Generation/1158/1). iFixit. . Retrieved June 25, 2010.
[51] "iPod Touch 1st Generation Teardown – Page 2" (http://www.ifixit.com/Teardown/iPod-Touch-1st-Generation/596/2). iFixit. . Retrieved October 4, 2009.
[52] "iPod Touch 2nd Generation Teardown – Page 2" (http://www.ifixit.com/Teardown/iPod-Touch-2nd-Generation/586/2). iFixit. . Retrieved October 4, 2009.
[53] "iPod touch 3rd Generation Teardown – Page 2" (http://www.ifixit.com/Teardown/iPod-touch-3rd-Generation/1158/2). iFixit. . Retrieved October 4, 2009.
[54] "IECEE – Public information. Model: 616-0404 DAP284864PA" (http://certificates.iecee.org/cbtestcert/cbtestcert.nsf/NCBPUB/0abf13f1341397b0c12574ab0023f32e). IEC. . Retrieved 2011-08-31.

# External links

- Official website (http://www.apple.com/ipodtouch)
- Technical Specifications (all models) (http://support.apple.com/specs/#ipodtouch)

# iPod Nano

Apple's 6th Generation iPod Nano

| **Developer** | Apple Inc. |
|---|---|
| **Manufacturer** | Foxconn |
| **Type** | Digital Audio Player / Portable Media Player |
| **Retail availability** | September 7, 2005–current |
| **Media** | Flash memory from 1 to 16 GB<br>(currently 8 and 16 GB) |
| **Operating system** | 1.3.1 (1st Generation)<br>1.1.3 (2nd Generation)<br>1.1.3 (3rd Generation)<br>1.0.4 (4th Generation)<br>1.0.2 (5th Generation)<br>1.2 (6th Generation) |
| **Power** | Lithium ion battery |
| **Display** | 176×132 1.5" color LCD (1G/2G)<br>320×240 2" color LCD 3G<br>240×320 2" color LCD 4G<br>240×376 2.25" color TFT-LCD 5G<br>240×240 1.5" color TFT-LCD 6G |
| **Input** | Click Wheel / Accelerometer / Multi-Touch |
| **Connectivity** | iPod 30-pin or 32-pin Dock Connector<br>USB 2.0 (wireless remote receiver)<br>3.5mm headphone jack |
| **Predecessor** | iPod Mini |
| **Related articles** | iPod Classic<br>iPod Shuffle<br>iPod Touch |

The **iPod Nano** (trademarked, marketed, and stylized as **iPod nano**) is a portable media player designed and marketed by Apple Inc. The first generation iPod Nano was introduced on September 7, 2005 as a replacement for the iPod Mini.[1] It uses flash memory for storage. The iPod Nano has gone through six models, or generations, since its introduction. The current "sixth generation" iPod Nano supports FM radio, a pedometer, and a 39.1-millimetre (**unknown operator: u'strong'** in) square 240×240 display with a multitouch interface.

## Supported audio formats

- AAC (8 to 320 kbit/s)
- Protected AAC (from the iTunes Store)
- MP3 (8 to 320 Kib/s, including variable bitrate files)
- Audible (formats 2, 3 and 4)
- Apple Lossless
- AIFF
- WAV
- MP4

## Development

Development work on the design of the iPod Nano started only nine months before its launch date.[2] The Nano was launched in two colors (black and white) with two available sizes: 2 GB (roughly 500 songs) and 4 GB (1000 songs).[1] On February 7, 2006, Apple updated the lineup with the 1 GB model (240 songs).[3] Apple also released some accessories, including armbands and silicone "tubes" designed to bring color to the Nano and protect it from scratches, as well as a combination lanyard-earphone accessory that hangs around the neck and avoids the problem of tangled earphone cords.

## Electronics

The iPod Nano uses general-purpose integrated circuits (IC) instead of smaller, low-cost custom-developed chips, possibly to reduce time-to-market. This design, however, increases the number of electronic components and increases the cost. Japanese engineers estimated the component cost of the 2 GB Nano as between JP¥ 22,000 and JP¥ 27,000 (US$ 185-US$ 227), which was high compared to the retail price of JP¥21,800 (US$183) at the time. The cost of 2 GB Nano flash memory was about JP¥14,000 (US$118). Apple also opted for the 0603 (1.6x0.8 mm) surface mount technology which was just beginning widespread use in mobile phones in 2005.[4] The iPod Nano uses a PortalPlayer PP5021C "system on a chip" with dual embedded 80 MHz ARM 7TDMI processors.[5]

## Reception

The initial consumer response to the iPod Nano was overwhelmingly positive and sales were heavy.[7] The Nano sold its first million units in only 17 days, helping Apple Inc. to a record billion-dollar profit in 2005.[8]

The first generation iPod Nano packaging.

Apple's release of the iPod Nano as a replacement for the iPod Mini was viewed by many as a risky move.[9] Steve Jobs argued that the iPod Nano was a necessary risk since competitors were beginning to catch up to the iPod Mini in terms of design and features, and believed the iPod Nano would prove to be even more popular and successful than the iPod Mini.

Within days of the Nano's release, some users reported damage to the Nano, suggesting that the LCD screen had become so scratched that it was unreadable, even when the backlight was on. Many have reported fine scratches on their Nano caused by microfiber cloths. Other owners reported that their Nano's screen cracked with no provocation.[10] On September 27, 2005, Apple confirmed a small percentage ("less than 1/10 of 1 percent") of iPod Nanos shipped with a faulty screen and agreed to replace any Nanos with cracked screens, but denied the iPod Nano was more susceptible to scratching than prior iPods.[11] Apple started shipping iPod Nanos with a protective sleeve to protect them from scratches. In October 2005 a class action lawsuit was filed against Apple, with the plaintiffs seeking reimbursement for the device, legal fees, and "unlawful or illegal profits" from sales of the iPod Nano. Lawyers for the plaintiffs claim that the devices "scratch excessively during normal usage, rendering the screen on the Nanos unreadable, and violating state consumer protection statutes".[12] Similar lawsuits were later filed in Mexico and the United Kingdom.[13] As of early 2009, Apple is in the process of settling a court case over the scratched iPod Nano screens, it has been suggested for Apple to set aside $22 million to refund users. A Judge will need to sign off the terms by April 28, 2009.[14] Some commentators such as BusinessWeek's Arik Hesseldahl have criticized the lawsuits. Hesseldahl dismissed them as "stupid" and suggested that they benefitted "no one but the trial lawyers," but also suggested that Apple could have avoided litigation by offering "full refunds on unwanted Nanos" instead of charging a re-stocking fee and lengthening the return period from 14 days (when purchased through Apple retail or online) to 30 or 60 days.[15]

The size of the package was reduced 50 percent with the introduction of the second generation.[6] The fourth and fifth generation packaging mimics this packaging, while the third generation used a larger but otherwise similar version of it.

## Incidents

In Australia, an iPod Nano caught fire while being charged on a PC.[16] [17] Since the limited warranty was expired, the consumer was not able to get a replacement right away.

Another iPod incident happened in the Hartsfield-Jackson Atlanta International Airport while a man was working in the airport.[18] His iPod Nano set his pants on fire. Apple Inc., refused to release a statement regarding this issue.

In addition, an iPod Nano sparked in Japan in January while it was recharging.[19] Although no one was injured during the incident, Apple Inc. investigated the incident.[20] It was reported on August 19, 2008 that 17 incidents of abnormal overheating with first-generation iPod Nano units while recharging had been reported in Japan, including cases in which *tatami* mats had been charred.[21] On August 10, 2010, Apple Japan released a statement saying that it would replace any iPod Nanos that overheated.[22] [23]

On November 11, 2011, Apple announced the iPod nano (1st generation) Replacement Program, intended to address concerns over overheating batteries.[24] Customers with affected devices can fill out a claim form to confirm eligibility for replacement. Defective devices will be replaced within six weeks and will carry 90-day warranties. Customers who have personalized iPod nano devices will not be able to receive personalization on their replacement devices.[25] During the replacement process, there have been several reports of users receiving an iPod nano 6th generation as replacement instead of the expected 1st generation device that users sent in during the recall.[26] Because using the iPod nano 6th generation with a Mac computer requires iTunes 10 or higher, which in turn requires Mac OS X Leopard system software, Apple will upgrade the system software of participants running earlier versions of OS X, on request.

## History

### First generation

Black first-generation iPod Nano.

On September 7, 2005, Apple introduced the iPod Nano at a media event with Steve Jobs pointing to the small watch pocket in his jeans and asking, "Ever wonder what this pocket is for?"[27] Advertising emphasized the iPod Nano's small size: 40 millimetres (**unknown operator: u'strong'** in) wide, 90 millimetres (**unknown operator: u'strong'** in) long, 6.9 millimetres (**unknown operator: u'strong'** in) thick and weighs 42 grams (**unknown operator: u'strong'** oz), its stated a battery life of up to 14 hours, and its screen of 176×132 pixels, 38 millimetres (**unknown operator: u'strong'** in) diagonal, displaying 65,536 colors (16-bit color).[28]

### Second generation

On September 12, 2006, Apple updated the Nano line. The second-generation Nano features scratch-resistant, anodized aluminium casing like the earlier Mini's design; the multiple color choices (silver, green, pink, blue, and black) mirror those of the Mini as well. However, unlike the second-generation Mini, the button labels do not match the color of the Nano. Instead, they are gray, like the first-generation Mini, except for the black iPod which has a black click wheel. The second-generation Nano features "a brighter, more vibrant display",[29] a battery life upgrade (from 14 to 24 hours), and doubled storage sizes with the new 2, 4, and 8 GB models (compared to the previous 1, 2, and 4 GB models). The second generation iPod Nanos also support gapless playback of audio files, a new search option, and a 40 percent brighter screen. The 2 GB model was available in silver only. The 4 GB was initially available in green, blue, silver, or pink. The 8 GB model was initially only available in black but Product Red was later added. Apple claims that the second generation iPod Nano's packaging is "32% lighter and uses 52% less volume than the first generation",[30] thereby reducing environmental impact and shipping cost at the same time.

On October 13, 2006, Apple announced a special edition iPod Nano Product Red, with a red exterior and 4 GB of storage. For each red iPod Nano sold in the United States, Apple donates US$10 to the Product Red initiative, while retaining the regular price.[31] On November 3, 2006, Apple introduced a red 8 GB model, due to "outstanding customer demand", while also retaining the same price point of the black model with an equally large storage capacity.[32]

On December 26–27, 2006, Apple's website and servers crashed due to thousands of people downloading iTunes software since so many iPods were sold that Christmas season.

### Third generation

A black 8 GB third generation iPod Nano.

Apple updated the Nano again on September 5, 2007. The third-generation Nano featured a 2-inch (**unknown operator: u'strong'** mm) QVGA (320 x 240) screen and a shorter, wider, heavier design, with new colors. New features included browsing via Cover Flow, a new user interface, video playback, and support for new iPod Games. Users had to repurchase games bought a month before the debut of the new iPod as they were not supported. The Nano was announced in a 4 GB version coming in silver and an 8 GB version coming in silver, turquoise, mint green, black, and Product Red. The battery lasted for approx. 24 hours on audio playback and approx. 5 hours on video playback. On January 22, 2008, Apple released a pink version of the 8 GB iPod Nano.

Combining elements from previous generations of the iPod Nano, the third-generation Nano had an aluminum front plate and a stainless steel back plate. The Nano also sported a new Minimalistic hold switch, similar to the iPod Shuffle's power switch, which had been moved to the bottom of the player. The 2-inch (**unknown operator: u'strong'** mm) screen had the smallest dot pitch of any Apple product, having the same pixel count as the 2.5-inch (**unknown operator: u'strong'** mm) display of the iPod Classic.

On October 6, 2007, Apple released a firmware update (1.0.2) via iTunes that was said to improve Cover Flow and yield faster menu navigation.[33] The update was also released for the iPod Classic. On November 28, 2007, Apple released another firmware update (1.0.3) via iTunes, which included unspecified bugfixes. January 15, 2008 saw the release of version 1.1, which added support for iTunes movie rentals, music song lyrics support and included more unspecified bugfixes. Apple released update version 1.1.2 in May 2008 and version 1.1.3 in July 2008 with even more bug fixes.

## Fourth generation

At the Apple Let's Rock Event on September 9, 2008, the iPod Nano Fourth Generation was officially announced.[34] It returned to the narrow form factor of the 1st and 2nd Generation model, while retaining and rotating the 51-millimetre (**unknown operator: u'strong'** in) screen from the 3rd gen model. It was also thinner than the 1G, 2G and 3G, measuring 90.7 millimetres (**unknown operator: u'strong'** in) tall by 38.7 millimetres (**unknown operator: u'strong'** in) wide by 6.2 millimetres (**unknown operator: u'strong'** in) thick, and weighing 36.8 grams (**unknown operator: u'strong'** oz). It had a curved aluminum shell and glass screen (the glass screen being held in place with nothing but the shell). Apple claimed the battery would last 24 hours of music playback, and only 4 hours of video playback, compared to the 5 hours of the previous generation.

Fourth generation iPod Nano.

The six previous colors (silver, black, mint, turquoise, berry red, and rose pink) were replaced by silver, black, purple, light blue, green, yellow, orange, red, and pink, for a total of nine, although the Product Red color was only available directly from Apple (website and retail stores). Apple marketed the new colors as "nano-chromatic". Also added was an accelerometer which allows the Nano to shuffle songs by shaking it, the option between portrait and landscape display modes by tilting the iPod left or right, and access to Cover Flow when tilted sideways.[35] Videos, however, could only be played in landscape mode. The user interface was also refreshed, adding a more stylized look in keeping with the new hardware design. It included a new voice recording feature which started automatically when an Apple compatible microphone is plugged in. It also included the new "Genius" feature, introduced by Apple the same day. The Genius feature automatically creates playlists based on a selected song using an algorithm built by Apple.

It was additionally touted as "the most environmentally friendly iPod Apple has ever made", containing arsenic-free glass and a BFR-, mercury-, and PVC-free design. It was also claimed to be highly recyclable. The iPod Nano 4G was shipped in cases similar to the 2G ones with the clear view in the front, and is marketed in three models: 4 GB (limited production to Europe only) and 8 GB and 16 GB. Limited quantities of an unannounced 4 GB model surfaced in various markets.[36] Also, the iPod Quiz game was dropped and replaced with a Maze game which makes use of the iPod's accelerometer similarly to such games on the iPhone and iPod Touch.

The fourth generation dropped support for charging via FireWire. "This change means that any dock accessories that use the dock connector's FireWire pins to send power—many older speakers and car chargers, for example—will not charge the 4G iPod Nano."[37]

## Fifth generation

At Apple's September 9, 2009 event, a fifth generation iPod Nano was unveiled with reduced prices on the larger model (at the time of release, the 8GB was priced at $149 and the 16GB at $179), a larger, 56.3 millimetres (**unknown operator: u'strong'** in) diagonal screen (up from 50.8 millimetres (**unknown operator: u'strong'** in) in third and fourth generation iPod Nanos), which is also wider,[38] integrated video camera with 16 special effects, microphone, FM Radio with iTunes tagging (via RDS) multiple radio regions including Americas, Asia, Australia, Europe, and Japan.

Fifth-generation iPod Nano camera and microphone.

It also includes Live Pause, a built-in pedometer, Nike+iPod Support and a speaker (features more anticipated with the iPod Touch).[38] [39]

The headphone jack and dock connector swapped locations so that the headphone jack is now to the left of the dock connector. Therefore the fifth generation iPod Nano uses a different Apple Universal Dock insert than the fourth generation.[40]

The fifth generation iPod Nano has nine finishes: Silver, Black, Purple, Blue, Yellow, Orange, Product Red, Green and Pink. All have a glossier, shinier finish than the fourth generation. Just like the 4G iPod Nano, Product Red and the yellow Nano were only available on the Apple Online Store and Apple Retail Store .

Its standard features include picture viewing and video playback. This model also has the Genius Mix feature installed.

## Sixth generation

At a media event on September 1, 2010, Apple announced the sixth generation iPod Nano, which among many new features, is designed around a high resolution square touch-screen.[41]

Sixth-generation iPod Nano.

The device now features a smaller 1.54 inch multi-touch screen with a lower resolution of 240×240 pixels but a higher pixel density of 220 pixels per inch,[42] as opposed to the larger 2.2 inch screen on the iPod Nano 5G. The device has a 0.39 watt-hour battery rated at 3.7 volts, giving a capacity of 105 mAh,[43] and specified to give 24 hours of music playback on a single charge. The device takes about 3 hours for one full charge. The device retains the same 30-pin dock connector as previous generations. The new iPod Nano has lost the previous generation's video camera, built-in voice recorder (although plugging in headphones with a built in microphone will reveal a Voice Memos app) and built-in speaker, and games. It has also lost support for video playback,[44] but music videos and video podcasts (vodcasts) can be synced onto the device, and the audio from them will play on the device, with a single key-frame shown on the screen.[45] It still includes the Nike+iPod fitness option as well as an FM radio tuner. It has a black on white screen contrast option and other accessibility options. The 6th generation iPod Nano has the same price point as the 5th generation device.

However, many consumers believed that the sixth generation was a downgrade, because it lacked some features that the previous versions had, such as the video camera, video playback, speakers and the click wheel, while also having a smaller screen.

A firmware update (version 1.1) for the Nano was released on February 28, 2011. The update added the ability to change songs or pause with a double click of the sleep/wake button. It also added the ability to turn the device off by holding the sleep/wake button. The user interface has also been enhanced. On October 4, 2011, the iPod Nano 1.2 update was unveiled at the Apple "Let's Talk iPhone" event at the Town Hall, 4 Infinite Loop. This update added the option to increase or decrease the size of the home buttons for easier use. The update also added a better fitness app, which has a better pedometer split into walking and running style. The update also included sixteen new clock faces, which include designs like a Nixie tube clock face or a old style clock face, and Disney licensed designs, such as Mickey Mouse and Kermit the Frog, bringing to a total of 18 clock faces. Three more background images were also added.[46] A watch band for the iPod Nano is available at the Apple store.

# Specifications

| Generation and Appearance | | Capacity | Colors | Connection | Original release date | Minimum OS to sync | Rated battery life (hours) | Screen (pixels) | Onboard RAM | Physical Size | Weight |
|---|---|---|---|---|---|---|---|---|---|---|---|
| 1 |  | 1 GB | Black White | USB (FireWire for charging only) | February 7, 2006 | Mac: 10.3.4 Windows: 2000[47] iTunes 5 or later | *Audio*: 14 *Slideshow*: 4 | 176×132 0.168 mm dot pitch | 32 MB | 89 mm 41 mm 6.9 mm | 42.5 g (1.5 oz) |
| | | 2 GB | | | September 7, 2005 | | | | | | |
| | | 4 GB | | | | | | | | | |
| | | *Replaced Mini.* Color screen for picture viewing; 1 GB version released later. | | | | | | | | | |
| 2 |  | 2 GB | Silver | USB (FireWire for charging only) | 12 September 2006 | Mac: 10.3.9 Windows: 2000[48] iTunes 7 or later | *Audio*: 24 *Slideshow*: 5 | 176×132 0.168 mm dot pitch. | 32 MB | 89 mm 41 mm 6.6 mm | 40 g (1.41 oz) |
| | | 4 GB | Silver Blue Green Pink | | | | | | | | |
| | | | (Product) Red Special Edition* | | October 13, 2006 | | | | | | |
| | | 8 GB | Black | | September 12, 2006 | | | | | | |
| | | | (Product) Red Special Edition* | | November 3, 2006 | | | | | | |
| | | Anodized aluminium casing with plastic top and bottom; 6 colors available. | | | | | | | | | |

<table>
<tr><td rowspan="4">3</td><td rowspan="4"></td><td>4 GB</td><td>Silver</td><td rowspan="3">USB<br>(FireWire<br>for<br>charging<br>only)</td><td rowspan="2">September<br>5, 2007</td><td rowspan="3">Mac: 10.4.8<br>Windows: XP<br>iTunes 7.4 or later</td><td rowspan="3">Audio: 24<br>Video: 5</td><td rowspan="3">320×240<br>204 ppi</td><td rowspan="3">32 MB</td><td rowspan="3">70 mm<br>52 mm<br>6.6 mm</td><td rowspan="3">49.3 g<br>(1.74 oz)</td></tr>
<tr><td rowspan="2">8 GB</td><td>Silver<br>Light<br>Blue<br>Light<br>Green<br>Black<br>(Product)<br>Red<br>Special<br>Edition*</td></tr>
<tr><td>Pink</td><td>January<br>22, 2008</td></tr>
<tr><td colspan="10">51 mm QVGA screen; lighter color shades and chrome back; new interface; video-playing capability.</td></tr>
<tr><td rowspan="3">4</td><td rowspan="3"></td><td>8 GB</td><td rowspan="2">Silver<br>Black<br>Purple<br>Blue<br>Green<br>Yellow<br>Orange<br>Pink<br>(Product)<br>Red<br>Special<br>Edition*</td><td rowspan="2">USB</td><td rowspan="2">September<br>9, 2008</td><td rowspan="2">Mac: 10.4.11<br>Windows: XP<br>iTunes 8 or later</td><td rowspan="2">Audio: 24<br>Video: 4[44]</td><td rowspan="2">240×320<br>204 ppi</td><td rowspan="2">32 MB</td><td rowspan="2">91 mm<br>38 mm<br>6.1 mm</td><td rowspan="2">36.8 g<br>(1.3 oz)</td></tr>
<tr><td>16 GB</td></tr>
<tr><td colspan="10">Curved enclosure and new colors; revised interface; voice recording features; "shake to shuffle"; accelerometer; limited 4 GB models[36]</td></tr>
<tr><td rowspan="3">5</td><td rowspan="3"></td><td>8 GB</td><td rowspan="2">Silver<br>Black<br>Purple<br>Blue<br>Green<br>Yellow*<br>Orange<br>Pink<br>(Product)<br>Red<br>Special<br>Edition*</td><td rowspan="2">USB</td><td rowspan="2">September<br>13, 2009</td><td rowspan="2">Mac: 10.4.11<br>Windows: XP<br>iTunes 9 or later</td><td rowspan="2">Audio: 24<br>Video: 5[44]</td><td rowspan="2">240×376<br>204 ppi<br>0.3<br>megapixel<br>camera</td><td rowspan="2">64 MB</td><td rowspan="2">91 mm<br>38 mm<br>6.1 mm</td><td rowspan="2">36.3 g<br>(1.28 oz)</td></tr>
<tr><td>16 GB</td></tr>
<tr><td colspan="10">Polished aluminium case including a larger screen, video camera, FM radio tuner, Recorder and a pedometer. Retains entire color line as fourth generation.</td></tr>
</table>

| 6 |  | 8 GB | Silver<br>Graphite<br>Blue<br>Green<br>Orange<br>Pink<br>(Product) Red<br>Special Edition* | USB | September 1, 2010 | Mac: 10.5.8<br>Windows: XP<br>iTunes 10 or later | *Audio*: 24[44] | 240×240<br>220 ppi | 64 MB[49] | 37.5 mm<br>41 mm<br>8.78 mm | 21.1 g<br>(0.74 oz) |
|---|---|---|---|---|---|---|---|---|---|---|---|
| | | 16 GB | | | | | | | | | |
| | | Multi-touch screen. No click-wheel, camera or video playback. The 1.1 OS update brought the ability of turning off by holding the wake/sleep button. Same price range as the 5th generation, except in Europe, Japan and Australia. Features iOS-like interface design and still contains "shake to shuffle", FM radio, and pedometer. The 1.2 OS update added built-in accelerometer support which works with Nike+iPod without the need to attach a Nike receiver or shoe sensor.[50]<br>* – Apple Store exclusive | | | | | | | | | |

## Timeline of compact iPod models

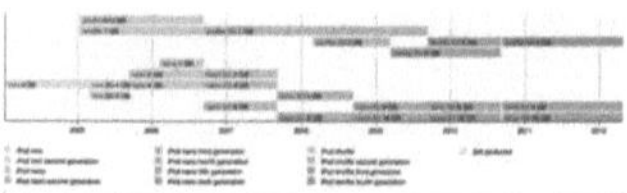

*Sources: Apple press release library,*[51] *Mactracker Apple Inc. model database*[52]

## See also

- Nike+iPod
- iTunes
- iPhone
- iPod Touch
- iPod Mini

## References

### Technical specifications

- First generation [53]
- Second generation [54]
- Third generation [55]
- Fourth generation [56]
- Fifth generation [57]
- Sixth generation [58]

## Footnotes

[1] "Apple Introduces iPod Nano" (http://www.apple.com/pr/library/2005/sep/07ipodnano.html) (Press release). Apple Inc. September 7, 2005. . Retrieved December 23, 2006.

[2] Grossman, Lev (September 12, 2005). "Stevie's Little Wonder" (http://www.time.com/time/magazine/article/0,9171,1103572,00.html). *Time*. . Retrieved October 3, 2006.

[3] "Apple Unveils New 1 GB iPod Nano at $149" (http://www.apple.com/pr/library/2006/feb/07ipod.html). *Apple press release*. Apple Inc.. February 7, 2006. . Retrieved December 23, 2006.

[4] Hirotaka Itoh; Naoki Asakawa (December 2005). "Looking Inside iPod nano, Game Boy Micro" (http://techon.nikkeibp.co.jp/article/HONSHI/20061122/124204/). *Nikkei Electronics Asia*. . Retrieved July 19, 2007.

[5] EveryMac.com. "Apple iPod nano 1GB, 2GB, 4GB Specs (MA004LL/A*)" (http://www.everymac.com/systems/apple/consumer_electronics/stats/ipod_nano.html). . Retrieved October 18, 2007.

[6] "Apple and the Environment" (http://www.apple.com/environment/design/). *Apple.com*. Apple Inc.. . Retrieved October 22, 2006.

[7] Michelle Myers (September 10, 2005). "Apple store buzzes with Nano fever" (http://news.com.com/2100-1041_3-5858457.html). *CNet*. . Retrieved September 15, 2005.

[8] Daniel Drew Turner (October 11, 2005). "Apple Hits $1 billion in Profit for 2005" (http://www.eweek.com/article2/0,1895,1870027,00.asp). *eWeek*. . Retrieved January 3, 2006.

[9] Burrows, Peter (September 13, 2005). "iPod nano: Will small size mean big sales?" (http://www.businessweek.com/the_thread/techbeat/archives/2005/09/ipod_nano_will.html?campaign_id=rss_blog_techbeat). *BusinessWeek*. .

[10] Angell, Larry (September 25, 2005). "iPod nano screen complaints abound" (http://ilounge.com/index.php/news/comments/ipod-Nano-screen-complaints-abound/). *iLounge*. . Retrieved September 27, 2005.

[11] Jim Dalrymple (September 27, 2005). "Apple responds to iPod Nano screen concerns" (http://www.macworld.com/news/2005/09/27/nanoscreen/index.php). *Macworld*. . Retrieved September 28, 2005.

[12] Kasper Jade; Katie Marshal (October 21, 2005). "iPod Nano owners sue Apple over screen issues" (http://www.appleinsider.com/article.php?id=1340). *AppleInsider*. . Retrieved October 21, 2005.

[13] "Nano lawsuit goes international" (http://news.bbc.co.uk/2/hi/technology/4417344.stm). *BBC News*. November 8, 2005. . Retrieved May 10, 2006.

[14] "Apple Settles iPod Nano Scratch Lawsuit with $25 Refund" (http://www.macrumors.com/2009/01/23/apple-settles-iphone-nano-scratch-lawsuit-with-25-refund/). Macrumors.com. January 23, 2009. . Retrieved November 29, 2011.

[15] Hesseldahl, Arik (October 27, 2005). "iPod nano Lawsuits: Who Wins?" (http://www.businessweek.com/technology/content/oct2005/tc20051027_665544.htm). *BusinessWeek*. . Retrieved November 8, 2005.

[16] Exploding iPod dies gruesome death (http://www.engadget.com/2007/04/12/exploding-ipod-dies-gruesome-death/), Engadget.

[17] Gadgets: iPod nano Explodes During Charge (http://gizmodo.com/gadgets/gadgets/ipod-Nano-explodes-during-charge-248859.php), Gizmodo.

[18] "iPod Sets Man's Pants On Fire – News Story – WSB Atlanta" (http://www.wsbtv.com/news/14271878/detail.html). Wsbtv.com. October 5, 2007. . Retrieved November 29, 2011.

[19] "iPod nano Sparks Investigation in Japan – GridLock – Just another KM / Tech Blog" (http://arjunthomas.com/?p=331). Arjunthomas.com. . Retrieved November 29, 2011.

[20] iPod nano emits sparks, Japan's government says (http://news.zdnet.com/2110-9584_22-6234041.html), Tech News on ZDNet.

[21] ???iPod?????????17? (17 overheating incidents involving old iPods while recharging) (http://headlines.yahoo.co.jp/videonews/jnn/20080819/20080819-00000051-jnn-bus_all.html), JNN News, (August 19, 2008). Retrieved on August 19, 2008. **(Japanese)**

[22] "Apple says to replace overheating iPods in Japan" (http://www.reuters.com/article/idUSTRE67A1DL20100811). Reuters. August 11, 2010. . Retrieved September 2, 2010.

[23] Koh, Yoree (August 12, 2010). "Apple Japan to Swap iPods that Overheat" (http://online.wsj.com/article/SB10001424052748704901104575423140720329642.html). *The Wall Street Journal*. . Retrieved September 2, 2010.

[24] Apple – Support – iPod nano Replacement Program (http://www.apple.com/support/ipodnano_replacement/), page on Apple Support site

[25] Apple Launches Worldwide Replacement Program for First-Generation iPod Nano (http://www.macrumors.com/2011/11/12/apple-launches-worldwide-replacement-program-for-first-generation-ipod-nano/) – Mac Rumors

[26] Joe Pollicino (December 21, 2011). "Apple reportedly swapping recalled iPod nanos for 6G counterparts; you have yours?" (http://www.engadget.com/2011/12/20/apple-reportedly-swapping-recalled-ipod-nanos-for-six-gen-counte). Engadget. . Retrieved December 21, 2011.

[27] Pachal, Peter (October 8, 2011). "Remembering Steve Jobs: His Best Keynote Moments" (http://www.pcmag.com/article2/0,2817,2394360,00.asp#fbid=lET7kiUDyS3). PCMag.com. . Retrieved November 15, 2011.

[28] iPod Nano (1st Generation) Technical Speifications (http://support.apple.com/kb/sp42) – Apple.com

[29] "Apple Introduces the New iPod Nano" (http://www.apple.com/pr/library/2006/sep/12nano.html). *Apple press release*. Apple Inc.. September 12, 2006. . Retrieved December 18, 2006.

[30] "Apple – Environment" (http://www.apple.com/environment/). Apple.com. . Retrieved November 29, 2011.

[31] "Apple Announces iPod Nano (PRODUCT) RED Special Edition" (http://www.apple.com/pr/library/2006/oct/13red.html). *Apple press release*. Apple Inc.. October 13, 2006. . Retrieved January 13, 2007.

[32] "Apple Announces New 8GB Model of iPod Nano (PRODUCT) RED Special Edition" (http://www.apple.com/pr/library/2006/nov/03nano.html). *Apple press release*. Apple Inc.. November 3, 2006. . Retrieved January 7, 2007.
[33] "iPod Classic, Nano updated to 1.0.2" (http://www.macnn.com/articles/07/10/06/ipod.classic.nano.102/). *MacNN*. October 6, 2007. . Retrieved January 22, 2008.
[34] Ricker, Thomas (September 9, 2008). "Official: iPod nano reaches 4G, looks tall for its age" (http://www.engadget.com/2008/09/09/official-ipod-nano-reaches-4g-looks-tall-for-its-age/). *Engadget*. . Retrieved September 12, 2008.
[35] "Identifying the 4th Generation iPod nano" (http://ipod.about.com/od/understandingipodmodels/g/4g_nano.htm). About.com. . Retrieved April 29, 2009.
[36] Topolsky, Joshua (September 17, 2008). "Official : Mysterious new 4GB iPod nano 4Gs begin appearing on store shelves" (http://www.engadget.com/2008/09/17/mysterious-new-4gb-ipod-nano-4gs-begin-appearing-on-store-shelve/). *Engadget*. . Retrieved September 17, 2008.
[37] Frakes, Dan (September 16, 2008). "Fourth-generation iPod Nano" (http://www.pcworld.com/article/151113/fourthgeneration_ipod_nano.html). *PC World*. .
[38] "iPod nano new features" (http://www.apple.com/ipodnano/features/). Apple.com. . Retrieved November 29, 2011.
[39] "iTunes – Partner with iTunes" (http://www.apple.com/itunes/tagging/). Apple. . Retrieved September 2, 2010.
[40] "Learn about Apple Universal Dock" (http://support.apple.com/kb/HT1380). Support.apple.com. April 22, 2010. . Retrieved September 2, 2010.
[41] "Apple unveils new TV box for renting movies, shows – Yahoo! News" (http://news.yahoo.com/s/ap/20100901/ap_on_hi_te/us_tec_apple). News.yahoo.com. . Retrieved September 2, 2010.
[42] "iFixit – iPod Nano 6th Generation Teardown: Page 1" (http://www.ifixit.com/Teardown/iPod-Nano-6th-Generation-Teardown/3563/1). iFixit. September 8, 2010. . Retrieved September 11, 2010.
[43] "iFixit – iPod Nano 6th Generation Teardown: Page 2" (http://www.ifixit.com/Teardown/iPod-Nano-6th-Generation-Teardown/3563/2). iFixit. September 8, 2010. . Retrieved September 11, 2010.
[44] "iPod nano – Read the technical specifications" (http://www.apple.com/ipodnano/specs.html). Apple. . Retrieved September 2, 2010.
[45] Starrett, Charles (September 9, 2010). "iLounge.com – iPod nano 6G plays music videos, vodcasts as audio" (http://www.ilounge.com/index.php/news/comments/ipod-nano-6g-plays-music-videos-vodcasts-as-audio/). iLounge.com. . Retrieved September 11, 2010.
[46] Horwitz, Jeremy (October 4, 2011). "New details on iPod nano 6G software 1.2" (http://www.ilounge.com/index.php/news/comments/new-details-on-ipod-nano-6g-software-1.2/). iLounge.com. . Retrieved October 4, 2011.
[47] "Apple Introduces iPod nano" (http://www.apple.com/pr/library/2005/sep/07ipodnano.html). Apple Inc. September 7, 2005. . Retrieved March 1, 2010.
[48] "Apple Introduces the New iPod nano" (http://www.apple.com/pr/library/2006/sep/12nano.html). Apple Inc. September 12, 2006. . Retrieved March 1, 2010.
[49] Rassweiler, Andrew (September 27, 2010). "iSuppli Estimates New iPod nano Bill of Materials at $43.73" (http://www.isuppli.com/Teardowns/News/Pages/iSuppli-Estimates-New-iPod-nano-Bill-of-Materials-at-$43-73.aspx). iSuppli.com. . Retrieved September 16, 2011. ""The most expensive subsection of the new nano is the memory system, featuring ... 512Mbits of mobile Double Data Rate (DDR) DRAM from Samsung""
[50] "Fitness Updated" (http://www.apple.com/ipodnano/features.html#fitness). Apple.com. . Retrieved November 29, 2011.
[51] Apple Inc., Apple press release library (http://www.apple.com/pr/library/), Retrieved September 19, 2007.
[52] Mactracker ( mactracker.ca (http://www.mactracker.ca/)), Apple Inc. model database, version as of 26 July 2007.
[53] http://support.apple.com/kb/SP42
[54] http://support.apple.com/kb/SP25
[55] http://support.apple.com/kb/SP14
[56] http://support.apple.com/kb/SP497
[57] http://support.apple.com/kb/SP569
[58] http://support.apple.com/kb/SP593

# External links

- Official website (http://http://www.apple.com/ipodnano/)

# Article Sources and Contributors

**SendStation_Systems** *Source*: http://en.wikipedia.org/w/index.php?title=SendStation_Systems *Contributors*: D6, Dr. Woo, Ryan Postlethwaite, 6 anonymous edits

**iPod** *Source*: http://en.wikipedia.org/w/index.php?title=IPod *Contributors*: -Edwin-, -Majestic-, .telus., 000peter, 100110100, 1337 pirate 13, 14545, 16@r, 16bitz, 1Goofy1, 2030a, 2100 x, 21655, 25or6to4, 2irbuland, 34i65278345, 3x0, 666iii, 68DANNY2, 91at1, 999999user, A Man In Black, A bit iffy, A. B., A3u2s8a5, A8UDI, AKemmler, AMK1211, ASG1022, AV3000, Aaron Brenneman, Aaron Schulz, Abadone, Abdullais4u, Abhiramesh, Abscissa, Academic Challenger, Acalamari, Acaz93, Ace of Spades, Aceofskies05, Acewolf359, Acps110, Acquire16, Acroterion, Acs4b, Adam Conover, Adam1213, Adambiswanger1, Adambro, Adamr20, Adashiel, Addict 2006, Addshore, Adjusting, Admiralthrawn 1, Adraeus, Adriaan, Adrian, Aeons, Afed, Aflorent1, Afterthought.cjb.cc, Agateller, Ageekgal, AgentPeppermint, Ahmad0410, Ahmadr, Ahoerstemeier, Aidan W, Aido2002, Airconswitch, Airplaneman, Aitias, Aj-pcfe, Ajcfreak, Ajfweb, Akamad, Akcarver, Akendall, Akkar, Akshayvarma89, Aksi great, Alahan, Alainr345, AlanArthurJenkins, Alansohn, Alasdair, Alaz, Albanaco, Alcoholica1, Alcuin, Aldie, Alecao89, Alegoo92, Alethiareg, Alex9865, AlexOvShaolin, Alexburke, Alexhchen, Alexignatiou, Alexisamattlover, Alfa989, Alfred Centauri, Ali@gwc.org.uk, Alibaba, AlinBoss, AlistairMcMillan, Aljurians, Allen3, Allstar86, Alphachimp, Alreajk, Alsandro, Although, Alton, Altzinn, Alu369, AluminumPancake, Amanda6758, Ambar, Ambarawa, Amberrock, Amicon, Amir Reza Moosavi, Amists, Ampedup, Andrejj, Andrevan, Andrew sh, Andrew1990, Andrewag, Andrewpmk, Andy120290, Andycjp, Andyroo316, Anecdote, Anetode, Angela, Angelic Wraith, Angelsfreeek, Angiiluvsfob, AngryParsley, Angrypriceandkobangus, Angus Lepper, Animal91X, Ank329, Anom8trw8, Anomalocaris, Anomalycp, Anon lynx, Anonymous Dissident, Anonymous editor, Anonymous1234567890, Anoojvedd, Ansett, Ant, Antandrus, Antialias, Antilived, Antonio Lopez, Anurag online, Aoi, Apedia, ApnAEA, ApolloBoy, Appl3zealot, Appleboy, April Arcus, Apsterling, AquaStreak, Aragami, Aralvarez, Arctic.gnome, Arda Xi, Argasp, Arghlookamonkey, Arjunm, Arkarian01, Armagion, Arminius, Arnavchaudhary, ArrowmanCoder, Art LaPella, Arthur Holland, Arthurbarnhouse, Artrix, Asafoetida, Asbestos, Asdavis, Asderoff, Ashsebs, Asim18, AtomicKitty2006, Attilios, AubreyEllenShomo, Auburn, AuburnPilot, Aude, Auiricle, AuldReekie, Aurelien Fache, AussieLegend, Austin Hair, Avaarga, Avicennasis, Avnit, Ayao1, Az1568, AzaToth, Azureman, B0gmen0t, BD2412, BJ Nemeth, BSKBGM7172495, BShiplet, Baa, Babajobu, Bad Camaro, Badcoverversion, Baka toroi, Ballin20914, Bandan, Bandaria, BanyanTree, Bardofcornish, Barefootguru, Barneca, Barneyboo, Barraki, Bartfat, Bartimaeus, Baryonic Being, Bayerischermann, Bbatsell, Bdesham, Bearian, BenKnauss, BenS, Bender235, Benwildeboer, Benzamin, Besidesamiracle, Bestsoft666, Betakate, Bewildebeast, Bfluid, Bh3u4m, Bhutti, Big Jock Knew, Big balls ben 28, BigBrightStars, BigCow, BigMar992, Bigbadbyte, Bigbluefish, Bigdon128, Biggoggs, Bigman241, Bigtimepeace, Biker14, Bingobangobongoboo, Bishonen, Bk0, Bkell, Bkkasper, Bknauss, Black Kite, Blackened, Blainster, BlaiseFEgan, Blakegripling ph, Blaske, Blorg, Blue520, BlueMint, Bluemilker, Bmader, Bmicomp, Bneely, Bobblehead, Bobblewik, Bobbyi, Bobianite, Bobmarly777, Bobo192, Bollinger, Bombcar, Bongwarrior, Bonkers bassist, Bonusballs, Bopash4, Boredzo, Bornhj, Bourega, Bowman, Boxstaa, BradBeattie, Bradley0110, Bramo126, Brandon10092, Brdforallseasons, Brendanfernandes, Breseler, Brian Kendig, Brian Mitchell, Brian0918, Brianpekol, Brianski, Brickwall2, Brighterorange, Brisvegas, BrokenSegue, BrowserWARwarning, Bruce010479, Brucethemoose, Bryantm3, Bssc81, Btipling, Btm, Bubba73, Buchanan-Hermit, Buckdog05, Bucketsofg, Bucksfan51391, Bucs, Buffalo242424, BuickCenturyDriver, Bull-Doser, Burn N Flare, Burner0718, Burningchrome, Buschaga, Bushing, Butterfly0fdoom, Bw1470, Bxj, Bysisky, C S, CHawke, CL, CLW, CMC, CMD Beaker, CNicol, CTU Kyoto, CWY2190, CWenger, CWii, Cactus.man, CalPaterson, CalendarWatcher, CalumCook234, Calvin 1998, CambridgeBayWeather, Can't sleep, clown will eat me, Canadian-Bacon, CanadianLinuxUser, Canar, Canberra photographer, CanisRufus, Canley, Captainstarbucs, Cardioscan, CaseyAlicia, Casper2k3, CatherS, Cathlicmetal, Cathryn J., Catotac, Causa sui, Cavrdg, Cbrown1023, Ccscott, Cdr1122334455, Cedars, Celerityfm, Celestra, CesarB, Cfawcett, Chabuk, Chairboy, ChangChienFu, Chappinator, Chappy84, Char645, Charles the chinese, Charlie2fart, CharlotteWebb, Chconline, Cheesy123456789, Chelsea Cantone, Chelseat04, Chemturion, CherryL.ollipops, Chiefskittles, Chitoboy, Chmod007, Choalbaton, Chochopk, Chopsticks, Chris Roy, Chris as I am Chris, Chris the speller, Chris1219, Chris3145, ChrisGriswold, ChrisLawrence2007, Chrisadamsa, Chrisca123, ChristTrekker, Christopher Parham, Christopher Thomas, Chych, Ciceronl, Cioxx, Ciphergoth, Citicat, Citikiwi, Ck lostsword, Ckatz, Ckelsh, Clam0p, Classicalgas, Classicfilms, Cleared as filed, Clem cowsie, Clever chonny, Cliff smith, Cliffb, ClonedPickle, Cmarcus, Cmdrjameson, Cmerca678, Cobaltbluetony, Cocoaguy, Coffee4binky, Coffeezombie, Cohesion, Coldheart95, ColinCameron, Collinhotchkiss, Colonies Chris, ColourBurst, Combuchan, Commander, CommonsDelinker, Compaq1011, Computerjoe, Condem, Conquerist, Conscious, Cookiehead, Cool Hand Luke, Cool-guy357, CoolFox, Cooljuno411, Coralys, Corevette, Corktin, Cornbread2145, Corpx, Cowfear, Cpfan0446, Craigbruce, Crashbandicoot, Crazeguy, Crazy Chimp, Crazycomputers, Crazytales, Crazyviolinist, Cream horn, Creidieki, Cremepuff222, Crissov, CritCol, Csaag, Csaribay, Csichris, Cskorik, Ctjf83, CurlyGirl93, Curps, Cwbh, CyberSkull, Cyberevil, Cybersayer, Cynical, CyrilB, Cyrus Andiron, D, DCM101, DDRFRK, DDima, DJGrieve, DKqwerty, DMCer, DMacks, DMurphy, DREWcorp, DVD R W, Daa89563, Dachande, DaedalusBTJ, Daev, Dale Arnett, Dali, Dallen19, Damieng, Damon Mah, Dan.cameron, DanTheMan702, DanWis, Dandanmatthews, Dane2007, Dangman4ever, Daniel5127, DanielCD, Danieldanielnova, Dankessler, Dannychoo, Dannydream9999, Dante Alighieri, Dantheman1128, Daredevil2794, Darijoe, DarkAudit, DarkSideMoon, Darkguy, Darkkefka, Darrenhusted, Darryliscool, Das7002, Dasani, Dashboardy, Dasler, Daunrealist, Dave souza, Davewild, David Fuchs, David Gale, David.Monniaux, David349, Davken1102, Dawnseeker2000, Dcandeto, Dck7777, Dcoetzee, Dcooper, DeVante', DeadEyeArrow, DeathShot39, Debresser, Decora, Deepa87, Defenestrate, Deffydeffydeffy, Defparadox, Dekisugi, Deli nk, Delldot, DeltaFalcon, Deltabeignet, Demosthenes 1, Dennin, Dennx, DerHexer, Design junkie, Desposy, Deusfaux, Devalleric, Deviantmind, Devoindahouse, Dgies, Dgurka052, Dgw, Dharmabum420, Di.ckFa.ceC.aptainFart, Dicconb, Dicklyon, Digitalme, Dillio411, Dim386, Dimwell, Dina, Dinomite, Discospinster, Dishwasherrat, Disk Crasher, Djbroox, Djcederberg, Djmckee1, Djyang, Dmitri Yuriev, Dmr2, DoSoMeThiN', Dockingman, Domthedude001, Donbert, Donfbreed, Doom127, Doublex120, Doug Bell, DougJustDoug, Doughboy92, Doulos Christos, Dowgy177, Dposse, Dr. Woo, DraQue Star, Dragon.slayer72, DragonWR12LB, Dragonfire75, DragonflySixtyseven, Dream out loud, Drizz 360, DropDeadGorgias, Drumguy8800, Drummyo, Dsdukcy, Dtcdthingy, Duckbill, Duke33, Duncan McB, Dunnybrusher, Dusoft, Dustinr88, Dylpickleh8, Dynaflow, Dysepsion, Dysprosia, Déjà Vu, EEPROM Eagle, ERcheck, ESkog, Eagle1.eagle1, Earlwayne, East718, Eatrains, Ebaci, Ed Poor, Ed g2s, Edd69, Eddiettr, Edgar181, Editcount tester, Edmunds, Edward, Edward Nygma, Edwy, Eekerz, Efficacious, Egglord, Ehurtley, Ekozie, Ekrub-ntyh, ElDarric, ElKevbo, Electricmoose, Electron9, Eleuther, ElfMage, Ellmist, Elmiko, Elsendero, Elvey, Elvis7hockey7, Embrio, Emericaskateforever, Emfraser, EmiOfBrie, EmperorOfSevenSeas, Empoor, Emre D., Epbr123, Ephilei, Epson291, Eptin, Equazcion, Eraserhead1, Erasmus16, Erasmus17, Erciesielski, Ergosteur, Eric343, Ericchubb, Ericlewis91, Ericnoel, Erik9, Errorlines, Esanchez7587, Esebi95, Esperant, Esprit15d, Etaoin, Eternal Pink, Euchiasmus, Euku, Evannxmcrr, Evanwil02, Evercat, EverettColdwell, Everyking, Everything counts, Evil Monkey, Evil saltine, Evosoho, Exert, Exeunt, Expectations, Extraordinary Machine, Exult, Exvicious, Eykanal, FF2010, FFAFRoxorzMyBoxorz, FIshstick, FOARP, Fabrictramp, Fairway, Falcon9x5, Family von stennett, Fandangojones, Faradayplank, Farcaster, Favonian, Feco, Feezo, Feffbiffmaff, Feinoha, FelisLeo, FellowWikipedian, Femto, FeralDruid, Ferdiaob, Ffiti, Finlay McWalter, Fireaxe888, Fireball1244, Firien, FisherQueen, Flai92, Flaming, Flamingspinach, Flarn2006, Flashsoccer2703, FletcherD, Flockmeal, Flowerparty, Fluence, Flyingostrich, Flávio Paiva F1, Fnfd, Foolfromhell, Footballboy12, Footballplayr69, Footoo, Forgetaboutit3000, Formina Sage, Foshizzel, Fourthords, Foxshox, Fr usc, Fr0oze91, Fram, Frankk74, Frazzydee, Freakofnurture, Fred Bradstadt, Free Bear, Freebie info, FrenchIsAwesome, FreplySpang, Frijole, Fritz Saalfeld, Frogzorz, FrostyBytes, Froth, Frozenfaith, Frungi, Frzl, Fshepinc, Fsotrain09, Func, FunnyYetTasty, Funnybunny, Funpage, Furrybeagle, Furrykef, Fuzheado, Fvw, Fxg97873, GMcdellan, GRAHAMUK, GRevolution824, GSAViolist, Gaborgasko, Gadgetguru1, Gaius Cornelius, Gallagher783, Garas, GarconDansLeNoir, Garrett Albright, Garrett18, Gary King, Gary Kirk, Gastel, Gateman1997, GatesPlusPlus, Gatta, Gaul, Gavani, Gazpacho, Gbelk08, Gblaz, Gdo01, Geeoharee, Gene Nygaard, Geneffects, GeneralDuke, Geni, Geoffspear, Geoking66, Geomguy, Georgalis, George1028, Gerbrant, Gesiwuj, Ghent, Gherb9, Ghingo, Ghoshsanjoy, Ghosthost, Giant89, Giantgrawp, Gigarice, Gilliam, Gilly0824, Ginsengbomb, Gjd001, Gladmax, Glen, Glenn Wolsey, Glennforever, Glennfoster, Glennwells, Gnowor, Gobonobo, Godlord2, Gogo Dodo, Goheels681, GoldDragon, Golden14, Goldfish007, Gomada, Gomm, Goober-peas, Googleaseerch, Goosemiester, Gordonjay, Gorillazx1, GraemeL, Graham87, Gralo, Grammarmonger, Grand River, Grandmastergrawp, Grant146, Grawp, GrayFullbuster, Greenday21, Greg Godwin, GregAsche, GregorB, Gregsimon, GreyCat, GrigorIII, Gringer, Grm wnr, Grstain, Grunt, Gscshoyru, Gtdp, Guanaco, Guanxi, Guitarrhythm, Gunit5000, Gunmetal Angel, Gurch, Gurchzilla, Gurubrahma, Gusapple, Gwernol, Gyrferret, Gywst, Gzkn, HIGH44576I, HINSTAMAN, Habanero-tan, Hack-Man, Hackerssidekick, Hadal, Hadley, Haha7778, Hailey C. Shannon, HalfShadow, Halfdan, Hamhurricane, Hamish2k, Hammer1980, Hananeko, Hannu, HappyCamper, Harambee, Hardyplants, Harlanh001, Harrison V, HarrisonB, Harry491, Harryboyles, Hasannur, Haukurth, Hawkdude1, Hayden120, Hchoi, Hcs, Hedgey42, Heimstern, Hellisp, Hello32020, Hello5959us, Hellogirls12, Henry Flower, Henry W. Schmitt, Henrygb, Herb-Sewell, HereToHelp, Hereford, Hgrosser, Hihihi124, Hippie Metalhead, Hirokagei, Hlclark78, Hoary, Hoho, Holeshot07, Hondaman, Honeyman, Honeymanl, Honsiong chong, Hootentogger, Hooverbag, Hoshie, Hotlorp, Howabout1, Howard Dawber, Howcheng, Htmlguy716, Huangcjz, Huggle, Husky, Husond, Hut 8.5, Hutcher, Huwr, Hybee, Hydnjo, Hydrargyrum, Hyperfried, Hypo11, I.am.lost, IDuDaChaCha, IE, IFEice, IKato, ILovePlankton, IMatthew, IRT.BMT.IND, ISquishy, Iamaditya, Ian01, Ian13, IanMcGreene, Ianblair23, Ianjones50, Iansexybeast, Iapetus, Iav, Iceberg3k, Iced Kola, Icedevil6, Idontlie52, Iediteverything, Ifny, Igoldste, Ijsportsman, Ikh, Ilia Kr., Ilikefood, Ilikepie2221, Illegal Operation, Illnab1024, Illspirit, Ilmari Karonen, IlyaHaykinson, Imamathwiz, Imaninjapirate, Imjake9, Imjustmatthew, Imrananwar, In23065, InShaneee, Indosauros, InfinityAndBeyond, Infioh, Infrogmation, Inkington, Inky, Inkypaws, Ino5hiro, Interiot, Intomwetrust, Invader2, Ipathshoes1985, Ipod guy, Ipodsocool, Ipodstabber, Irdepesca572, Iridescent, Iron Dragon91, Irregulargalaxies, Isilanes, Islander, Itself, Iuhkjhk87y678, Ivan Pozdeev, Ixfd64, Izhmal, J Di, J Dogg, J matula, J.delanoy, J0 13, J306, J3tolmer1, JBelkin, JDonkey123, JFG, JForget, JFreeman, JGT-5151, JGabbard, JHMM13, JHunterJ, JLaTondre, JMikeD, JNW, JOSamsung, JP Godfrey, JR98664, JYOuyang, Jab843, Jacarv, JackCarter, Jackgeddes1234, Jackwosere, JagSeal, Jakal 121, Jake Nelson, Jakg, Jalabi99, Jamcamuk, James086, JamesHPurdy, JamesR1701E, Jameswc, JamisonK, Janejellyroll, JanusK, Jaranda, Jared Hunt, Jason Stormchild, Jav72, Java13690, JaviC, JayKeaton, Jayleen23, Jazo123, Jcw69, Jdmt, Jdtyler, Jdude204, Jed meyers, Jeff G., Jeff8765, JeffStickney, Jeffgoz, Jeffr, Jenny Wong, Jeremy Visser, JeremyA, JesseGarrett, Jfwjxn, Jgjg123, Jh51681, Jherm, Jhessela, Jhsounds, JiFish, Jiang, Jic, Jim.henderson, Jimbo Wales, Jimeree, Jimmayoy007, Jimmy Green1, Jimothytrotter, Jimscott09, Jjdamaan, Jjman2010, Jjnguy, Jkeller007, Jmorrisseyd, Jmrowland, Jo Alem, Jo3.scott, JoanneB, Jodamiller, Joe99sam, JoeBlogsDord, JoeSmack, Joebloggsy, Joedadog2, Joel.labes, Joemc72, Joey Penguine, JoeyEleff, Jogloran, Johan Elisson, Johan h, Johann Wolfgang, John Fader, John0027, JohnJSal, Johnburnham192, Johnjosephbachir, Johnleemk, JohnnyBGood, Johnpseudo, Johnteslade, Jojhutton, Jollybengali, Jon Stockton, JonHarder, JonathanDP81, Jonel, Jonghyun0411, Jonnyx13, Jonrev, Jonv112, Jonwilliamsl, Jookieapc, Joranrose, JordeeBec, Jorge22s, JorgeGG, Jorophose, Jose209, Joseph11h, Joshbrez, Joshschr, Joshspazjosh, Jossi, Joxy, Joyous!, Jp3z, Jpatokal, Jpgordon, Jpittman, Jpk, Jpo, Jpp42, Jsfouche, Jstar8, Jtalledo, Jtdude101, Jteazy, Jtkiefer, Jtnt, Juckum, Julian Grybowski, Juliancolton, Jungleman1, Jurriaan van Hengel, JustAddPeter, JustinHall, Justinemery, Justinxtreme, Justme89, Jw12321, Jw21, Jwink3101, Jwissick, Jwood74, K.lee, KAMiKAZOW, KB1KOI, KUsam, Kablammo, Kafziel, Kalicaloo1212, Kamens, KamuiShirou, Kareeser, Karimarie, Karl Dickman, Kaszeta, Katalaveno, Kathleen.wright5, Katieh5584, Kawika, Kayotic, Kaysov, Kbdank71, Kbolino, Kc0wys, Keepitreal74, Keepsleeping, Keldan, Kencaesi, Kent Wang, Kerberus, Kesh, Kevinharbin, Kf4bdy, Khalid hassani, Khoikhoi, Kibbled bits, Kickassinator23, Kickin' Da Speaker, Kidfromnowhere, Kigoe, Kilrothi, Kiltedspartan, Kimchi.sg, Kimhill, King Cole, King of Hearts, Kingj123, Kipijo42, KirbyMeister, Kirin, Kiscica, Kiteinthewind, Kittyslasher, Kjkj17, Kkailas, Klhuillier, Kliqjaw, Kmg90, KnowledgeOfSelf, Kntrabssi, Koavf, Koman90, Komusou, Kondspi, Konman72, Konstable, Konstantino, Kookcooza, Koopa turtle, Kornfan71, Korske, KowDude, Kozuch, KrakatoaKatie, Krasimir, Krazycev13, Krellis, Krich, Kristmace, Kristof vt, Kross, Krun, Krystyn Dominik, Kubigula, Kukini, Kungfuazn, Kungming2, Kurt Jansson, Kuru, Kusma, Kusunose, Kylevk, Kyoko, Kyouteki, Kyro, Kythias, L Trezise, L0uls3, LAAFan, LDHan, LFaraone, LHOON, LOL, Laalaax3, Lababidi, LabradorLover456, Lakhim, Lamename3000, Landroo, Lankiveil, Lar, Larsinio, Latka, Laug, Laurens, Lawrencekhoo, Lc2022, Leaderofearth, LeeHunter, Leeborkman, Leithp, Len115566,

Lenin and McCarthy, Leonthes, Lethalwiki, LewisJeandren, Lewys93, Lexi Marie, Lexischemen, Lexusdevotee, LeyteWolfer, Lhasapso, Liene, LifeStar, Light current, Lightdarkness, Lightmouse, Lightsup55, Ligulem, LikeLakers2, Lilac Soul, Lillie666, Limulus, Lindosland, Linkspamremover, Linnell, Liquidspirit, LittleOldMe, Littlebr96, Liverpool Scouse, Lizardnada, Llamagirl625, Lo2u, Localzuk, LockeShocke, LocoBurger, Loft, Loggie4, Lol, Longhair, Loodog, Lord Of The Crayons, LordBleen, Lordkazan, Love2sppooggee, Lovely Chris, LovesMacs, Lowellian, Lrose551, Lt. penguin, Lubos, Lucid, Luckas Blade, Ludraman, Luk, Luna Santin, Lurker, Lw99dds, Lynx46, M1ss1ontomars2k4, M3tal H3ad, MBCF, MBK004, MC MasterChef, MECU, MER-C, MFago, MJisnotmylover, MKoltnow, MOCKBA, MONGO, MP3Additions, MPD01605, MZMcBride, Maartenvdbent, MacGyverMagic, MacPrince, Macaddct1984, Macic7, MacintoshWriter, MadFunk, Madchester, Madeinusa, Madhero88, Magic100, Magicmarc, Magikdonut, Magister Mathematicae, Magnificator, Magnus Manske, Mahanga, Mailer diablo, Majoreditor, Majorkev, Majorly, Makeyourself, Makmaxima, Malber, Malleus Fatuorum, Malo, Mamaluigibob, MamboJambo, Mandarax, Mandy55566ui, Mangojuice, Mani1, Manvillez, Maquirri91, Marcham89, Marcika, Marco255, Mareino, Marek69, Mark Dingemanse, Mark Heiden, Mark83, MarkSweep, Markintellect, Markmack0604, Marknagel, Markremo, Markt3, Martarius, Martha p, Martin451, MartinRobinson, Martinp23, Martinribadeneira, Master of Puppets, Mastercheif, Masterhand10, Matharvest, Matt Whyndham, Matt Yohe, Matt1992, MattTM, MatthewMastracci, Matticus78, Mattwolf7, Mauimonica, Maury Markowitz, Max.goedjen, Maxamegalon2000, Maxim, Maxster, Maxx113087, Maybemaybe, Mb1000, Mbatman72, Mbessey, Mboverload, Mcamargo, Mckoss, Mcmillin24, Mcwetboy, Mczack26, Mdarwin2003, Mdfillipp, Mechanical Animals 666, Mee Merone, Meeples, MegX, Megahurtz83, Megamanfan3, Member, Memo232, Mendaliv, Mentifisto, Mercury12, Mereda, Messatsu, Metagraph, Metalmallow, MetraB, Mhking, Michael miceli, Michaelas10, Michaelkourlas, MichaelsProgramming, Michaelseko, Mick Duffys Spare Shack, Midkay, Mike 7, Mike Rosoft, Mike1, MikeWren, Mikegrant, Mikeryz, Mikesolo, Miles, MilitantRabbit, Mindspillage, MindstormsKid, Minesweeper, Minimac93, Minna Sora no Shita, Mintchocicecream, Mipadi, Miracleworker5263, Miranda, Mirv, Mishmash8, Mister macphisto, Mistercow, Misty1, Misza13, Mixmatch, Mj 1903, Mjmarshall, Mkid4500, Mlewan, Mo0, Mobius, Mobovivo, ModernTenshi04, Modify, Moe Epsilon, Moeron, Mojo-chan, Mojo29, Molteanu, Mongo95, Monkeyman, Mono, Moocow12345, Moocow135, Moogle10000, Moomoomoo, Moon unit 3000, Moooon8, Moptophaha, Mordomo, Moreno Valley User, Moriori, Morte, MosheA, MosheZadka, Motboylol, Mpeisenbr, Mpiff, Mproud, Mr Adequate, Mr.Do!, Mr.Xp, Mr.Z-man, MrBeast, MrMurph101, Mrdonnelly, Mrg5283, Mrmaroon25, Mrmerino, Mrmikenator, Mrmiscellanious, Mrqueen, Mrschimpf, Mrtrblmkr1, Mshonle, Mslake, Mspraveen, Mteorman, Mudcower, Muddypuppy17, Mugunth Kumar, Muhand, Mulad, Murphykidash, Murr, Muscort, Mushin, Mushroom, Music Sorter, MusicAndMath, Musicpvm, Mustafa Al Habib, Muéro, Mwakin21, Mx3, Mxn, MxyzptlkD5, MySchizoBuddy, MykReeve, Myscrnnm, Mysekurity, Mywymedia, N1h1l, N328KF, N3X15, N3m6, NHJG, NICKISYOURRULER, NJA, NMChico24, NSLE, NSR, Naalberg, Nagy, Nahallac Silverwinds, Nakon, Nalvage, Nandesuka, Nanouk, NastalgicCam, Natalie Erin, Nathan lethal, Nathan8225, Nathanziarek, Navito, NawlinWiki, Nbarth, Neale Monks, Ned Scott, NeilN, NekoFever, NeoVampTrunks, Neonblue2, Neonrusso, NerdyScienceDude, Neschek, Netsnipe, Neverquick, NewYorkDreams, Newguineafan, Ngapleaz, Nibuod, Nich148 9, Nicholas spurway, Nick, Nick R, Nick123, Nickavl, Nickelbacknut, Nickshanks, Nicolas1981, Nicoli nicolivich, NiggardlyNorm, Nigholith, Nightscream, Nihiltres, Nihonjoe, Nikosapi, Nilfanion, Nils, Nintendude, Nintenman, Niteowlneils, Nixdorf, Nixeagle, Nixón, Nizamarain, Nneonneo, No Guru, Nohat, Noiseactivist, Norm, NorwegianBlue, NotMuchToSay, NpaK13, Nscheffey, Nuiop729, Number 0, Nunh-huh, Nurmsook, O mores, OMGsplosion, ONEder Boy, Oagersnap, Oatmeal batman, Oberst, ObeyK1NGTaz, Obli, Obocat, Odicsan, Oe1kenobi, Off!, Oh.cera, Ohconfucius, Ohnoitsjamie, Ohthelameness, Ojibberish, Ojw, Oli Filth, Oliver Yee, Olorin28, Olvert, Omar6000, Omegatron, Omex, Omfghax, Omicron91, Omicronpersei8, OnBeyondZebrax, Onyiexca1mill, OrbitOne, OregonD00d, OriginalGamer, OriginalJunglist, Oriolpont, Oroso, Oscarthecat, OtherPerson, Outriggr, OwenX, P. Romack, P.B. Pilhet, P0per, PH1RESTRIKE, PHenry, PJM, PMHauge, PPGMD, PS2pcGAMER, PS3 FTW, PTSE, Pabouk, Page2, Panicciac, Panser Born, ParisianBlade, Parklandspanaway, Patchy1, Patricia62893, PatrickKnight, Patrickmcgrath, Patstuart, Paul August, Paul Haymon, Paul1337, PaulHanson, Pauly04, Paxsimius, Pb30, Pbui44, Pcarbonn, Pdefer, Peak Freak, Pearljam123, Pecholobo, Pedant17, Peipei, Pejhman, Pele Merengue, Penguinpablo, Pepper, Pepperpocket, Perardi, Perfectblue97, Perfecto, Peripitus, Persian Poet Gal, Peruvianllama, Peter, Peter Brockie, Peter S., Peterburk, Peteturtle, Petho, Petra87, Petri Krohn, Pevalou, Pewwer42, Pezza Mac, Pgk, Pgr94, Phenry, PhilBob Smells, PhilHibbs, Philip Pasch, Philip Trueman, Philippe, Philwelch, Phoebe, Phoenix2, Photographerguy, Phr, Pi, Piano non troppo, Pieleric, Pikaco, Pilgore, Pilotguy, PinchasC, Pixelperfect777, Plastictv, Plastikspork, Plaztikstarz, Plop, Pmsyyz, PoccilScript, Podfather, Poeloq, Pol098, Polaris 2.9, Polarman..AaA, PoliticalJunkie, Polonius, Pomegranate, Pongo489, Pontificalibus, Poomccheese, Porkman Pork, Porqin, Portillo, Possum, Postdlf, PowerMacX, Ppk01, PranksterTurtle, Prateekrr, Prewrite1, PriceCullen, Prodego, Profoss, Promethean, Proofreader77, ProveIt, Prowikipedians, Psantora, Psychofox, Psyk0, Ptdecker, Pteren, PullTheWires, Pumapayam, PureLegend, Pyrrhus16, Qaovxtazypdl, Qazwsx1, Qtac, Qtpa2tnh, Quadell, Quickliger, Quintote, Quixotic Rick, Quuxplusone, Qwerty Binary, QwertyD, R'n'B, R'son-W, R3m0t, R4wr, RHB, RHaworth, Rachzach96, Racliv, RadRafe, RadicalBender, RadioActive, RadioFan, Radmad, Raekwon, Raggedtoad, Raidako, Rajiv Varma, Rajrajmarley, Ramallite, Randomfrenchie, Randomgbear, Randy Johnston, Ranunculoid, Ranveig, Raskolnikov4138, Rassul, RastaKins, Rathee, Rattlesnake, RattusMaximus, Raul654, RaviC, Ravzz, Razorflame, Rcandelori, Rch2005, Rdsmith4, Rebeccafong, Reco, RedWolf, Redeyes, Redjar, Rednblu, Redsparta, Reedmalloy, Reedy, Reeveorama, Rehcsif, Reisio, Reliableforever, Rellum, Remember the dot, Remstar, Repetition, Repo2, Retired username, RetroPod, Rettetast, Reub2000, Reuvenk, Revasser, ReviewDude, Revived, Revrit, RexMaztar, RexNL, ReyBrujo, Reywas92, Rhanyeia, Rhobite, Rhun, Riana, Ricardoboricua2, Ricardocolombia, Rich Farmbrough, Richard Allen, Richmeister, Ricky81682, RideABicycle, Rillian, Rjcflyer@aol.com, Rjd0060, Rjoseph, Rjwilmsi, Rmarvel, Rmhand, Rmsuperstar99, Roastytoast, Robchurch, Robert Harper, RobertG, Robin-april, Roccyraccoon, RockMFR, Rockgodkp, Rockysmile11, Rodrigo Cornejo, Roeme, Rofthorax, RogerRanger, Rogerd, Rogerthat, Rohan2kool, Romansanders, Romestar, Romney yw, Romtobbi, Ron Ritzman, RoninRVP, Ronline, Rookkey, Rorschach, Rory096, Ross Uber, Rosswf, Roux, Rover007TN, RoyBoy, Royalguard11, Rpvt, Rruelas, Rsharich, Rskjr, Rspeed, Rudykog, Rugby471, Rumple84, Run!, Runewiki777, Rurik, Rushed4nr, Ruszewski, Rwils, Rwl4, RxS, RyanGerbil10, Ryanbomber, Ryanbutterworth, Ryank808, Rydra Wong, Ryulong, Rzelnik, S1066, SCEhardt, SGDentarthurdent, SHCarter, SJP, SPARTEN II, SPBLU, SPT, SUND33P, SYSS Mouse, Saaga, Sabeen557, SagaciousAWB, Sagaciousuk, Sahaskatta, Salazan, Saltycrayons, SalvadorRodriguez, Sam Francis, Sam Korn, Sam Li, Samadam, Samir, SamirKagadkar, Sampi, Samsoncity, Samsquamptch, Samuel Blanning, SamuelGrauer, Sandgem Addict, Sandstein, SandyGeorgia, Sango123, Sardanaphalus, Saros136, Sarrica, Sasha Slutsker, Satori Son, Saucealexman, Sbcsbc, ScaldingHotSoup, Scepia, SchfiftyThree, SchmuckyTheCat, SchuminWeb, Schurginator, Schzmo, Scootey, Scottydude, Screwball23, Sdeeyre, Sdfisher, Sdkfdslkf, Sdluvzu4, Sdornan, Sdrtirs, Sean William, Seano1, Searchme, Seb az86556, Seb26, Selmo, Semcguire, Semidelicious, SereR0KR, Serge KGB, Sergiy.gagarin, Sertion, SexyBern, Sfacets, Sfarmer76, Sfgiants320, Sgs101, Shaddack, Shadotron, Shadow demon, Shadow1, Shadow2598, Shadowjams, Shadowmario64fan, ShakingSpirit, ShaneCavanaugh, Shanel, Shanes, Shas, Shawn21099, Shawnc, Shawnhath, Shayonsaleh, Shieber, Shimishimiman, Shinjiman, Shinmawa, Shinshin, Shitmaster 5, Shoeofdeath, Shreshth91, Shyking, Sidewithus, Sietse Snel, Sifaka, Silencer01, Silivrenion, Sillstaw, SimFan10076, Simeon H, Simmonsjd02, Simon Shek, SimonD, SimonMackay, SimonP, Simonlebon, Simsimtigger, SiobhanHansa, Sir Nicholas de Mimsy-Porpington, Sir Stig, Sir Vicious, SirGrant, Sitenl, Sitush, Sjakkalle, Sjenkins7000, Sjorford, SkeletorUK, Ski splat, SkiBumMSP, Skillz187, Skittlesrgood4u, Skizzik, Skopelos-slim, Skuld, SkyBoxx, SkyWalker, Skylercall, Skytreader, Slakr, SlapAyoda, Slickshooter001, Sloman, Slon02, Smartguy777, Smearp, Smnioffe, Smokeonit, Smyth, Snapplegal, Snoopyish 2b, Sockatume, Soldierboy06, Som1Udontknow, Someguy1221, Someguy432, Somnium, Sonny Lei, Sonucoolkid, Sooner Dave, Soonercary, Sooperfalcon8, Sotomura, Soundslikealotofhooplah, Soupmix, Southfield, Soyturk, Spangineer, SparhawkWiki, Sparklamb698, Special-T, Speck-Made, Spectacgurl215, Specter01010, Speedway, Speer320, Spiff, Spiff666, Splash, Sploo22, Spongesquid, Spudzonatron, Spunkmeyer, Spunkybailey, Squiggleslash, Srhuston, Srippon, Srw985, St ttb, St.daniel, Staffwaterboy, StanBrinkerhoff, Stayfree, Steel, Stephan Leeds, Stephen B Streater, Stephen Compall, Stephen Shaw, SteveSims, Steveklein, Steven X, Stewartadcock, Stewie814, Stickguy, Stirwen, Stizz, StrengthOfNations, Sturm105, Stwalkerster, SubwayEater, Sudoku4U, Suduser85, Suomi Finland 2009, Supapuerco, Super4814, SuperDude115, SuperJ9587, Supernabo, Supersaiyanplough, Surv1v4l1st, Sverdrup, Swaq, Swarve, Swatjester, Swid, SwordfishII, Syferus, Symode09, Synchroblst, Syvanen, Szayat, T Cash, T c951, T3h, TEG24601, TJ Spyke, TJDay, Tachiuchickoma, Tagishsimon, TakuyaMurata, Tamajared, Tamer of hope, Tangotango, Tareq Dowla, Tasc, Tawker, Taxman, Tbhotch, Tcncv, Tealwisp, Techjunky, Tedernst, Teenwriter, Teflon, TehBrandon, Telcik, TenOfAllTrades, TenPoundHammer, Terence, Terra Xin, Tesseran, TexasAndroid, TexasDex, Texture, Tgeairn, Thaaster, Thatdog, Thatoldsoul, The Anome, The Blue Lion, The Haunted Angel, The Hybrid, The Land of Smeg, The Man in Question, The Monster, The Noodle Incident, The Obento Musubi, The Orange, The Placebo Effect, The Rambling Man, The Rumour, The Spith, The Statue With The Dictionary, The Thing That Should Not Be, The Wookieepedian, The isteve, The wub, TheDudeAndAPen, TheEditrix, TheFreeloader, TheHYPO, TheNewPhobia, ThePinkEmoRanger, TheSiGuy, TheTruthiness, Thealok, Theaveng, Thebonobo, Theda, Thedjbogus, Thegriffin, Theimacguy, Thejackhmr, Thelb4, Thelockedown, Themindset, Theonlyedge, Theothernextdimension, Therampage, Theresa knott, Theropane3u483, Theseijiman, Theshadow27, Think outside the box, Thinspirit, Thirtysixway, Thiseye, Thistle172, Thnr, ThomasHarte, Thomasexciting, Thoraxcorp, Thrash25, Thue, Thumbsucker-UK, Thumperward, Tiddly Tom, Tide rolls, TigerShark, Tim1988 2, TimMorris, Timcruz, Timeforacatnap, Timl2k4, Timmcgraw, Timothylord, Timrem, Timro, Timsdad, Tinny186, Titan Rain, Titoxd, Tiucsib, Tjp1982, Tkelly888, Tkgd2007, Tkircher, Tlbrown4567, ToMmYgUn, Toad of Steel, Tobeyjaggle, Tobias Conradi, Tobyc75, Tobymarshman, Toddhansen, Tohru Honda13, Tokek, Tokinwalrus, Tom Herbert, Tomatoesrfruits, Tomhannen, Tomhormby, Tommyt, Tonei, Tony1, TonyTheTiger, Tooki, Toonmon2005, Toosmart215, Torolegend, Torritorri, TotoBaggins, Towel401, Toytoy, Tpbradbury, Trebor, Tree Biting Conspiracy, Treff89, Tregoweth, Trekky0623, Tresiden, Trevor MacInnis, TrevorLSciAct, Trevyn, TreyHarris, Trhaynes, Tri400, Trieste, Trisreed, Triver83, Trumpet marietta 45750, Tsunaminoai, Tsusurfanami, Tuspm, TutterMouse, Twang, Twinxor, Twister Twist, Two stripe, Twobells, Tydus Arandor, Tykell, Tymothy, Typhoon, Tyrhinis, U, UKWiki, UTSRelativity, UberScienceNerd, Ubergenius, Ucanlookitup, Ugen64, Uggh, Ukexpat, Ulmanor, Ulquiorra, Ultimate753, Ultimus, Ultras GE, Umanouski, Unam1292, Unbreakable MJ, Uncle G, Unforgettableid, UnicornTapestry, UnrealCoopNet, Urhixidur, UriBudnik, Uriah923, Urion2699, Userof2, Uthbrian, Uucp, V Man767, VSquared, VT hawkeye, Vafc4life, Valisol, Vanished user 5zariu3jisj0j4irj, Varnav, Vary, VashiDonsk, Vector Potential, Veemonkamiya, Vegasrebel29, VeiledAbyss, Vertigociel, Vexion321, Violetriga, ViolinGirl, ViperSnake151, Viridis, Vision Insider, Voltron, VolunteerFD, Voortle, Vter4life, WAZAAAA, Wackymacs, Wackyvorlon, WadeSimMiser, Wafulz, WakiMiko, Walkiped, Wall0159, Walloon, Wandering perfect fool, Wannger27, Warren, Wavemaster447, Wayward, Wcquidditch, Wdflake, Webdinger, Wehe, Welsh, WereWolf, Weregerbil, Wereon, West Brom 4ever, Wetyuppy, Weyes, Whatever you want it to be, WhisperToMe, Whomp, Whosyourjudas, Whywhenwhohow, Wiki alf, Wikieditor06, WikipedianMarlith, Wikiraphe, Wikiweb87, Wiknerd, Wild Bill, Wildcat dunny, Wilkus, Will866, William.li.1983, Willking1979, Willkm, Willyrules, Wilt, Windchaser, Windward1, Winner11111, Wisdom89, Wisq, Wj32, Wknight94, Wompa99, Wonk619, WoodlakeMan, Woohookitty, Wrockca, Wyatt th3 one ninja, X 0, X!, Xaosflux, Xcyuen, Xdannock, Xdenizen, Xeen, Xenctuary, Xenophrenic, Xevious, Xezbeth, Xiahou, Xinoph, Xizer, Xk, Xnatedawgx, Xxrobtasicxx, Y2kcrazyjoker, Ya2sin, Yacov, Yakuman, Yamamoto Ichiro, Yandman, Yanksox, Ybbor, Year 2144, YellowMonkey, Yidisheryid, Yoasif, Yoganate79, Your1727, Yourcool135, Yrrot, Yshoulduknow, Yuckfoo, Yuymako, Z.E.R.O., ZV1998, Zac and jordan, Zachlipton, Zakhebeone, Zambaccian, Zanimum, Zanorath, Zapptastic, ZayZayEM, Zenlax, Zenmark, Zenmaster137, Zero6532, Zeromaru, Zetawoof, Zfr, Zhou Yu, Zip22, Zippanova, Zippo, ZippyGoogle, Znalbone, Zoara, Zodon, Zoe, Zoeb, Zooobala, Zootsuits, Zorxd, Zsinj, Zumbojo, Zupahzmurv, Zweifel, ZyphBear, Zzedar, Zzuuzz, Zzyzx11, ~shuri, Ævar Arnfjörð Bjarmason, , , 4868 anonymous edits

**Macintosh** *Source*: http://en.wikipedia.org/w/index.php?title=Macintosh *Contributors*: 0pen$0urce, 1+2x3=9, 12.234.49.xxx, 209.75.42.xxx, 212.53.104.xxx, 21655, 24.15.135.xxx, 24.45.80.xxx, 2D, 32-bit dirty?, 3Coins, 4piecemcnugget, 63.192.137.xxx, 7, 75tpickupsx1983, 790314002, ?alabio, A bit iffy, A100nonymous, A2-computist, A8UDI, Aaron Schulz, Abce2, Abebenjoe, Abenjoel13, Abrech, Academic Challenger, Accurizer, Acps110, Acroterion, Adamdavid85, Adashiel, Adbmice, Addshore, Adjkabeef, AdjustShift, Adjusting, Admrboltz, Adz-834, Ae-a, Aerialvendetta, Aeryck89, Aetheling, Afknlnmfs, Ageekgal, Agendum, Ahoerstemeier, Aido2002, Airplaneman, Aitias, Akamad, Akcarver, Alainr345, Alansohn, AlbertR, Aleenf1, Alegoo92, Alex Heinz, Alex.muller, Alex22351, Alexf, Alexius08, AlexiusHoratius, Ali@gwc.org.uk, Alison, AlistairMcMillan, Allmightyduck, Allstarecho, Alpha Rex, Alphachimp, Altenmann, Althepal, Amatulic, Ameliorate!, Amyroe200, Anas 1762, AndersL, Andres, AndrewHowse, Andrewlp1991, Andrewpmk, Andy pyro, Andypandy.UK, Angela, Angelic Wraith, Angelsfreeek, Angr, AnonHat, Anonymous 198736, Anonymous 57, Anonymous 94056, Anonymous Paul, AnsonMBO, Anss123, Antandrus, Ante Aikio, Anthony, Antimac, Anville, Anvish,

Apparition11, Apple2gs, Applehockey87, Applesmellz, Appleuser4, AquaStreak, Aquietlittlecat, Arbero, ArchStanton69, Arcyf, AriX, Arjunpachory, ArnoldReinhold, Arrista30, Arru, Ash, Aspects, Astral, Astronautics, Asyndeton, Athomsfere, Atlant, Aude, AugieD369, AussieBoy, Austin Hair, Avicennasis, Avoided, Avs5221, Awesome.js, AxelBoldt, Ay98182, Az1568, AzaToth, B, Backmaybe, Baconcheeseburger, Bad Bad Guy, Bagatelle, Bajajvikram, Barntks, Barrylb, Baryonic Being, Bassplaya7, Bateau, Bathrobe, Batman tas, Batmanand, Bazymac, Bbatsell, Bbpen, Bbriggs1, Bdude, Beao, Beastlyskillzz, Beau-lane, Beefman, Beeswaxcandle, Belovedfreak, Ben Ben, Ben Boldt, Ben-Zin, BenBurch, Benscripps, Bento00, BetaGamma, Betta Splendens, Bettia, Bevo, Bfigura's puppy, Bhadani, Bheron, Bhny, BigBrightStars, BigBurkey, BigDunc, BigFatBuddha, BigHaz, Biglovinb, Bill PLates, BillFlis, Blainster, BlaiseFEgan, Blanchette, Blood sliver, Bloodshedder, BlueBeach, Bluedude5887, Blueking12, Bluhd, Bmdavll, Bob Schaefer, Bobblewik, Bobdobbs1723, Bobo192, Bobthechick, Boing! said Zebedee, Boner2k8, Bongwarrior, BonzaiRob, Booieh, Bordgious, Boredzo, BorgQueen, Borisavljevic, Borisborf, Bpazolli, Bradv, Brian Kendig, Brian Patrie, Brian Tvedt, Brianga, Bricology, Brighterorange, Brion VIBBER, Brmwk, Brooklynl, Brossow, Bruce1ee, Brucethemoose, Brusselsshrek, Bryan Derksen, Bryanfosterthesexy, Bte288, Bubba73, Bubbadaboy, Bullzeye, BurtPeck, Burzmali, Bus stop, BusterD, BusterD public, Butterfly0fdoom, C S, C-squared, C.Fred, C628, CWY2190, CWii, Cab88, Cachdis123, Caltas, Calvin 1998, Cam493, CambridgeBayWeather, Camhusmj38, Camw, Can't sleep, clown will eat me, CanadianLinuxUser, Canderson7, CanisRufus, Canthusus, Captain Wikify, Carsten Levin, Carter4444, Caseybutt, Casmith 789, Catdude, Catgut, Catslash, Causa sui, Cayzle, Cbh, Cdespinosa, Cdmarcus, Cedars, Cei, Cenarium, CenozoicEra, Centurion123, Centurion31337, Chaffers, Chainsawriot, Chairboy, Chardish, Charles Edwin Shipp, Charles Gaudette, Charleschuck, Charlie Quebec Delta Echo Seven Sierra Foxtrot, Chaser, Checco, Chemmer, Chick Bowen, Chicocarlucci, Chillin69x, Chmod007, Choalbaton, Chris 73, Chrislk02, Chrisrowhatey, ChristTrekker, Christopher Parham, Chriswiki, Chubbles, CimanyD, Cinnamon42, Ciphergoth, Cjmarsicano, Ckatz, Clam0p, Classicfilms, Cleared as filed, Clevelandmusic24, Cliffb, Closedmouth, Cmdrjameson, Cnjartist, Cobbb, Coleflinn, Colin13, Collinrocks26, Cometstyles, Commit charge, CommonsDelinker, Compilation finished successfully, Computerdan000, Connectionfailure, Conner 36, Connormah, Conversion script, CoolFox, Coooltonnn., Cosmo0, Courcelles, Cpchase, Cpl Syx, Craft Bio, Crazysim, Creeper1342, Cregq, Crem23, Cremepuff222, Cretin's Bezoar, Crohnie, Cryosheath, Csaag, Csl77, Ctachme, Ctjf83, Cureden, Curps, CyberSkull, Cybercobra, Cyberia23, Cyktsui, DARTH SIDIOUS 2, DF08, DGJM, DHeyward, DMacks, DSGPS, DSLITEDS, Da Dude, DaL33T, DabMachine, Dale Arnett, Dalek king7, DalekClock, Damian Yerrick, Damicatz, Danallen46, Danamania, Danielfolsom, DannyDaWriter, Dannydream9999, Danrha, Danstein7, Dantheman531, Darcagn, Dark Shikari, DarkElrad, Darks0uldier, Darth Panda, Darth aavon, Darwinek, Dave Farquhar, Dave souza, David Fuchs, David Wahler, DavidSJ, Davidm617617, Dawnseeker2000, Dbfirs, DeadEyeArrow, Debatewise, Debresser, Deconstructhis, Deflective, Dekisugi, Delldot, Deltabeignet, Denimadept, Deor, DerHexer, Deus Ex, DevastatorIIC, Devon rodriguez, Dewet, Di Stroppo, Diagonalfish, Diemunkiesdie, Digana, Digi23azlan, Dims110, DirtPriest, Discospinster, Diyar se, Dlohcierekim, Dmlandfair, Dmsar, Dnstest, DocWatson42, Docski35, DoctorW, Dogguitar, Dogsandcatslivingtogether, Dominic Hardstaff, Dominic7848, Donreed, Doovinator, Doradus, Doulos Christos, Douthy, Download, Dr Debug, Dr Fell, Dragomiloff, Dream of Goats, DreamGuy, Dsyzdek, Dtunnicliff, Durin, Dustan1456, Dylan620, DylanW, Dysprosia, Dzubint, E. Sn0 =31337=, EGroup, ESkog, EVula, Ebbers.ja, Ebe123, Edgar181, Edgarde, Editoruk, Edward, Edwy, Ehurtley, Eighth90, Eiscir, El C, El Krem, ElAmericano, Elf, ElinorD, Elkman, Elliotjordan, Ellmist, Elmo456, Elvey, Encephalon, Enchanter, Encyclopedant, Energetic is francine@yahoo.com, Enric Naval, Eo01415, Epastore, Epbr123, Equendil, Eran of Arcadia, Ergbert, Ergosteur, Ergzay, Ericoides, Erik9, Escape Orbit, EscapingLife, Esong200, EthanolRules, Evercat, Everyking, Everything counts, Exia, Eyeareveeyeen, Falcorian, Fan-1967, Fang Aili, Favonian, Fede.Campana, Felimrules, Feneeth of Borg, Fenevad, FeralDruid, Fern2131, FernandoAires, Fg2, Fieldday-sunday, Figureskatingfan, Final88, Finlay McWalter, Fire 55, Fireborn, Fishtorte, Fiskars007, Fjord23, Flcelloguy, Flewis, Flowerpotman, Fluffernutter, Flydpnkrtn, Foant, For great justice., Fortunateson, Fourthords, Frazzydee, Frencheigh, Frequencydip, Freshacconci, Freyr, Frijole, Frungi, Fshepinc, Fun and games are fun, Funandtrvl, Funnybunny, Fuzy2K, Fvw, Fæ, G-O-D of W-A-R, GGreeneVa, GHe, GRAHAMUK, GSMR, Gadha, Gah4, Gaia Octavia Agrippa, Gaius Cornelius, Gakusha, Galor, Galoubet, Gareth Owen, Garion96, Garrett Albright, Garrett247, Garrettw87, Garrithklingel, Gary King, Gateman1997, Gateway122, Gaz9407, Gblaz, Gdo01, Generica, Geoff.green, Geologist001, Geometry guy, GeorgeLouis, GeorgeStepanek, Gexmofia, Ghatten, Gholam, Gil Dawson, Gilliam, Glacialfox, Glen, Glen Pepicelli, Glenn, GnuTurbo, Gogo Dodo, GoldRenet, Goldfish007, Golftheman, Goncalves229, GoogledIt?, Gracenotes, GraemeL, Graemesmith, Graham87, GregAsche, GregPanos, Griffin5, Grm wnr, GroundedZero, Gscshoyru, Gsg9378, Gunshoters, Gurch, Gurubanks, Gus, Guy Harris, Guy M, Guyinblack25, Gwernol, Gyounk, Gyre, H92, HELLO 101, HJ Mitchell, Hahahaidontgetit, Haham hanuka, Haikupoet, HamburgerRadio, HammerinHank, Hanselcat, HappyCricketer, Happydude22, Harley22xxx, HarryHenryGebel, Harryboyles, Haseo9999, Hateless, Hayabusa future, Hayne, Hdlafglh, Hdt83, Headbomb, Hellboy4311, Hen7713, HenryLarsen, Hephaestos, Heptapod, HereToHelp, Heron, Hetar, HiddenInPlainSight, Hiddenpwner, Hillbrand, Hippo X, Hiroshimaboomboomboom, Hoho832, Holodoctor1, Honytawksroprater, Hoolme, Hootentogger, Hopping cat, Hotlorp, Hrajt, HubmaN, Human.v2.0, HummadJ, Husond, HuwPrestatyn, Hyad, Hydnjo, Hydrargyrum, Hydrogen Iodide, IAMAHIPO ocolor, IG Slayer, Iain.dalton, Iamalexjones77, Iameukarya, Iamunknown, Ian The Great, Ian13, Ianblair23, Ianweller, Ickyelf, Ignatzmice, Igoldste, Ijbond, IlliterateSage, Ilse@, Iluvcapra, Ilyaroz, Imperial92, Impfac, IndulgentReader, Ineffable3000, Infodriveway, Inks.LWC, Insolectual, Intelligentsium, Intersting, Ioeth, Iohannes Animosus, Iridescent, Irishguy, IronGargoyle, Irvela, IsUsername, IstvanWolf, Itfcsam, Ivymike21, Ixfd64, J.delanoy, JCarriker, JDP90, JForget, JNW, JP Godfrey, JR98664, JThinger, JaGa, Jaavaaguru, Jabencarsey, Jacek Kendysz, Jack The Hat, Jackalope darko, Jackfork, Jackmorino, Jackol, Jacobolus, JagSeal, Jakekito, Jamcamuk, Jamiemcc, Janizary, Janke, Janus Shadowsong, Jared, Jason C.K., Jasonq, Jaxl, Jcw69, Jdude204, Jeff G., Jeffrey Smith, Jeremiahhsu, JeremyA, Jerrie32, JhastyII, Jhertel, Jim Douglas, Jim Shaud, JimEOme, Jimthing, Jleedev, JoanneB, JoaoRicardo, JodyB, JoeCool59, Joelbelknap1, Joemexico16, Joemexico17, JoeyJoeyJoey, John Price, John Quincy Adding Machine, JohnInDC, JohnRDaily, JohnSawyer, Johnleemk, Johnmarkos, Johnrpenner, Jojhutton, Jokestress, Jonnyboy911, Jonpaulusa, Joost, Jordandanford, Jordanduhawesomedude, Josh the Nerd, Josh59x, JoshEdgar, Jossi, Jothesmo, Jotting9566, Jouer au Fat, Jovianeye, Jpk, Jsephton, Jtalledo, Jtbean, Jtkiefer, Jtn1209, Jtuttle, Julesd, Jusdafax, Jutiphan, Jw21, Jwissick, Jyoyopaek, Jéské Couriano, KGasso, KJS77, Kakaman, Kakurady, Kalathalan, Kaly99, Kanags, Karenjc, Karrmann, Kasskid, Katefan0, Kbolino, Kchishol1970, Keegan, Keilana, KelvSYC, KenshinWithNoise, Keremtezcan, Kesac, Kesh, Kevin chen2003, Keyser Söze, Kickapples, Kierenfahey, Kiko4564, Killthemac, King of Hearts, Kingpin13, Kiscica, Kissedbythesun, Kiteinthewind, Kjetil r, Klausness, Klilidiplomus, Knowhow, KnowledgeOfSelf, Knoxus, Koolkid2168, Korg, Kozuch, Kraetos, Krakeder9, Kross, Kubigula, KuduIO, Kukini, Kungfuadam, Kuru, Kwamikagami, Kyoko, Kyphe, Kyro, L Kensington, L3vyk3v4nd44l1, LAX, LBHS Cheerleader, La goutte de pluie, Laaaaabob, Lacrimosus, Lame Name, Lars T., Law, Ldo, LeaveSleaves, Leer25, Leftus, LegitimateAndEvenCompelling, LeilaniLad, Leon7, Leszek Jańczuk, Lethe, Lexein, Lfaporsche2424, Libstooge 01, Liftarn, Lightdarkness, Lightlowemon, Lightmouse, Lijorijo, Lilj242, Limelightist, LinaMishima, LindsayH, Linnell, LittleOldMe, Liubei, Llhm, Localzuk, Logan598, Loganm10, Lolcoaster, Lord Vader, LordBleen, Lostchicken, Lotta12, LovesMacs, Lowenddan, Lproven, Lradrama, LtPowers, Luigiwax, Luk, Lukobe, Luna Santin, Lupin, Luumu, M.nelson, ME4236, MER-C, MFNickster, MKoltnow, MONGO, MORRISPC10, MPD01605, MSTCrow, Ma8thew, Mabdul, Mac sux, Mac128, Mac2333, Macaddct1984, Maccomputerssuck, Macinjosh87, Macintosh User, Macintoshrocks9, Macrhino, Maebmij, MaganT2k13, Magicaldave, Magister Mathematicae, Mahanga, Mailseth, Makaristos, Makeemlighter, Makesdark, Malimbar04, Malo, ManIsWhiteMonkeyIsBrown, ManfrenjenStJohn, Manoflast3, Manufracture, Mareino, Marek69, Mario91234, Marj Tiefert, Mark5677, MarkGallagher, Markonen, Marskell, Martin451, Martinman11, MastCell, Master Freeman, Mathfan, Matt Britt, Matt the mexo, Matthew2414, Matthewwilliamproctor, Mattisse, Mav, Max Naylor, Maxim, Maxis ftw, Mboverload, McDonaldsGuy, McDoobAU93, McSly, Mcclurd, Mdd4696, Mdwh, Melmann, Mentifisto, Mephistophelian, Mera james, Merlion444, Mhkmack, Michael Hardy, Michael K. Edwards, Michael Snow, Michael93555, Michaelas10, Michaelbusch, Michelle eris, MidgleyDJ, MidiMacMan, Mihai1025, Mika1h, Mike Rosoft, Mike hayes, Mike1, Mikeo, Miles, Millifolium, Mimigu, Minesweeper, Mintchocicecream, Mipadi, Miquonranger03, Mirek2, Mirror Vax, Miss Morning, Missmarple, Misza13, Mkay222, Mklbtz, Moe Epsilon, Moncrief, Mono, Moondyne, Moonriddengirl, MorningView375, Morten, Mproud, MrC539, MrClean03, MrKalamazoo2, MrRedwood, Mrclark411, Mrholybrain, Ms ArtGeek, Msikma, Mtk402, Muchclag, Mushroom, Muzzy136, Mvelez0002, Mvjs, Mwk teck, Myanw, Mysterioususer, Mystiq0, Mystyc, Mütze, N3rf360, NHRHS2010, Nakon, Nanshu, NapoliRoma, Nappilainen, NathanHurst, Nauticashades, NawlinWiki, Ndenison, Ne1, Neil D, Neilfein, Neoleeyennek, Neonleon123456789, Nerdd, NerdyScienceDude, Nesnej15, Netizen, Netminder, Netoholic, Neverquick, NewEnglandYankee, Newkai, Nice Dad, Nice yum felicia@yahoo.com, Nickgill1, Nicoanison, Niemasd, Nigholith, Night of the Big Wind, Nightscream, Nikbro, Nikorasu, Nintendros, Nishkid64, Niteman555, Nixdorf, Nlu, Noetica, Nolat, NorseLord, NosraePekuL, NotSuspicious, Nothlit, Nricardo, Nsaa, Nsayer, NuclearWarfare, Nuggetboy, Numberp, Nwwaew, ONUnicorn, Oberonfoxie, Obli, Oblivious, Oda Mari, Ohconfucius, Ohnoitsjamie, Ohwhatcoolk, Okma, Oldschool, Oli-Was-Here, Olivier, Omar abdallah, Omegatron, Omicronpersei8, Omondra, One, Oneupthextraman, Onorem, Onthost, Onza110, Optimist on the run, Optowiki, OrangeCrusader, OrbitOne, Orbitalwow, Oregonerik, Ori, OriginalGamer, Osxshortcuts, OverlordQ, Owlmonkey, Oxymoron83, Oyvind, Ozguy89, PC destroyer, PJM, PR3V3NTION, PS3 FTW, Pablomartinez, PageVandal, PanamaDrummer911, Pantergraph, Papadopa, Paraded junk soul, Parklandstudent, Particleman24, Patrick T. Wynne, Patrick0Moran, Patton76, Paul Magnussen, Paul3144, PaulGS, Paulkman, Paulmrussell12, Paxsimius, Payco, Pbroks13, Pdk, Peama058, Pedro, Peeka, Pekaje, Penfold1000, Permafrost, Persian Poet Gal, Petesmiles, Petri Krohn, Pgk, Phaedriel, Phantomboy98043, Pharaoh of the Wizards, Pharos, Phdrummer, PhilZ, Philipdw, Philiscool1919, Philosophicles, Philwelch, PhoenixVTam, Phuk you1, Physman, Pidgeot, Piledhigheranddeeper, Pill, Pilotguy, Pineapple1, Pinkadelica, Pjoef, Planetary Chaos, Pmurph, Po panda, PoccilScript, Pol098, Polarbowler, Polarit, PoliticalJunkie, Poopshnicles, Poplggp, Porqin, Possum, Praetorian42, Prari, Primate, Prodego, ProhibitOnions, Prolog, Proud, Psantora, Pschemp, PseudoSudo, Psychoman247, Ptaylor270782, Puji6, Purinflash, Pyryle, Python eggs, Pyxzer, Qinnyc, QofASpiewak, Quadell, Quantpole, Quuxplusone, Qwe, Qwerty1211, Qwertyiscool1234567, Qxz, R'n'B, R.carroll, RAMChYLD, RCpatroler, RadicalBender, Raffaele Megabyte, Raidon Kane, RainbowOfLight, Ramallite, Rambo23, Ramunnoodle, Ramunnoodles, Ramurf, RandomStringOfCharacters, RandomUsefulKnowledge, Randum, Rattlesnake, RattusMaximus, Raul654, RaviC, Rbj, Rcandelori, Rchamberlain, Rdikeman, Rdsmith4, Reaper Eternal, Rebroad, Red, RedHillian, RedSpruce, RedWolf, Reddi, Redeming, Redgreen88, Reedy, Registeraccount, Reisio, Remember the dot, Remstar, Repetition, Res2216firestar, Retsoperedevlebrm, Revolution221, RexNL, Rich Farmbrough, Richard D. LeCour, Richardcavell, Rigel1, Rita Moritan, Riverwaste, Rjairam, Rjwilmsi, Rklawton, Rnbangras, Robb0082, Robert Brockway, Robert P. O'Shea, RobertG, RobertMfromLI, Rodhullandemu, Rogerd, Romansanders, Ronhjones, Ronk01, Room01, Rory096, Roryethanr, RossPatterson, Rossami, Rostz, RoyBoy, Rpresser, Rrburke, Rschen7754, Rsholmes, Rugby471, Rx 100e, RxS, Ryan Norton, RyanCross, RyanGFilm, Ryanaxp, Ryguy999, Ryrivard, S, SCJohnson77, SD6-Agent, SMC, ST47, Saalstin, Safarirss6394, Salanus, SallyForth123, Salty!, Sam Korn, Sandridge, SandyGeorgia, Sango123, Saptahrshi, Sardur, Sarranduin, Save-Me-Oprah, Scanlan, Scepia, Sceptre, SchfiftyThree, Schissel, Schmidtmandaddy, SchnitzelMannGreek, School99144, Schumi555, SchuminWeb, Schutz, Scientus, Sciurinæ, ScottSteiner, ScottlovesFanta, Scrotie, Scrumshus, Scwlong, Scythre, Sdedeo, SeaFox, Sebapeluca, Seidenstud, Seinfreak37, SenorKristobbal, Seth ze, Severusskad, Shadowlynk, ShaneKing, Shanel, Shaneo95, Shanes, Shanesan, Shannon Bradley, Shenme, Shinjiman, Shirik, Shizane, Shoeofdeath, Shoy, Simeobindo, Simmonsjd02, Simpsonfan25, Singapore Jones, Sionus, Sir Anon, Sir Nicholas de Mimsy-Porpington, Sir Stig, SirJibby, Skier Dude, Skunkboy74, Skyezx, Slady, Slakr, Slark, Sleigh, Slugger, Smalljim, Smyth, Snigbrook, Snowolf, SoCalSuperEagle, Sobrow, SolFruit, Somebody in the WWW, Someguy1221, Something12356, Sophisticated 23, Sora089, Soumyasch, Sp, SpaceFlight89, SpaceRocket, Spacesoftware, Spartan-James, Spencerperry, SpigotMap, Splash, SpookyMulder, Spritzsinger, SpuriousQ, Spyzy, Srikeit, Stan Shebs, Staney23, Starnestommy, Statsong, Steekel, Stefan, Stephan Leeds, Stephen Shaw, Stephenb, SteveRamone, SteveSims, Stevefis, Stevejobsforpresident, Steven Zhang, Stevenj, Stevenrasnick, StrawBerryBannaCream, Sturm55, Subsailor, Suffusion of Yellow, Suhailpeerbhai, Summitstudent, SuperHamster, Superdude876, Surv1v4l1st, Sverdrup, Switchfoot, Symode09, SyntaxError55, Syrthiss, SystemBuilder, TBSchemer, TDS, TEG24601, THEN WHO WAS PHONE?, TWCarlson, Takuzinis, Tangentidea, Tango, Tangotango, Taroaldo, Tarquin, Tassedethe, Tau'olunga, Tawker, Taxman, Tbhotch, Tbutz, Tcncv, Tdog106106, Teacish, Tech30, Techpodusa, Teles, Tempshill, TenPoundHammer, Terrillja, TexasAndroid, Tgeairn, Thakur rigved, The Catfish, The Cunctator, The Gilly, The Grate Spanish Duck, The Interior, The Land of Smeg, The Mysterious El Willstro, The Parting Glass, The Rambling Man, The Thing That Should Not Be, TheDJ, TheDoctorZoidberg, TheSheriff2004, Theanthrope, Theblueslime, Thechoirlife2006, Thehelpfulone, Thejadefalcon, Thejaggedtimes, Thesab, Thespencerjones, Thewallowmaker, Thg250, Thingg, Thomas Ash, Thomas Gilling, ThomasHarte, ThomasK, Thortful, Thursdayx20, Tide rolls, TigerK 69, TigerShark, Tim1357, Timrollpickering, Titoxd, Tktktk, Tman8809, Tmarki, Tobacco bean, Toddan, Tom Morris, Tommy2010, Tony1, TonyClarke, Tooki, Toonmon2005, Tooza64, Torbix, Tosha, TotalSpaceshipGuy3, Totie, Toussaint,

Toytoy, Tpbradbury, Tragen1234, Trailbum, Travelbird, Treaeakjdsfaew4h'iiioii, Tree Falling In The Forest, Trevor MacInnis, TrevorLSciAct, Trickybiz, Triwbe, Troy 07, Truenaruto, Trusilver, Turkeybacon97, Tutnkmn, Tuxisuau, Tverbeek, Twang, TwilligToves, Tyler, Typhoon, Tyw7, Ubcule, UberPwnage92, UchihaKazu, Ufosjt, Ulric1313, Ultra Galahad, UncleBubba, UncleDouggie, Unreal7, Unyoyega, Upi, Usb10, Usgnus, Uzsa, V-train, VMS Mosaic, Valermos, Validation7, Vanderbeekfan13917, Vary, VegKilla, VegaDark, Veinor, Veldin963, Velella, Venom098, Versus22, VibhuC, Vikramc84, Vizcarra, Vranak, Vrenator, Vulpes vulpes, WJSProwler, Wackymacs, Wakleon, Wampa Jabba, Wapcaplet, Warren, Wartrakk, Wavelength, Waycool27, Wayward, Wcquidditch, Webeditor909, Wechselstrom, Wendell, Wereon, Wernher, Wertjoe, West Brom 4ever, West.andrew.g, Wgungfu, Whale, Whitebox, Whywhenwhohow, Wifione, Wiki Roxor, Wiki alf, Wiki b00b12, Wiki man120, WikiED, WikiLaurent, WikiLegend12, Wikimassiker, Wikipedian456, Wiknerd, WillMcC, William Avery, Wimt, Win2000Pro, WindowzRULZlolZ, Wine Guy, Winston007, Wintersun777, Wipsenade, Wjmummert, Wknight94, Woodsstock, Woodwynlane, Woofs, Wpedzich, Wrcmills, Wtmitchell, Ww2censor, X!, X570, Xaosflux, Xdriver206, Xelentoliver, Xeworlebi, Xiggle, XuperUber, Yakbasser, Yaksar, Yamen993, Yamla, Yankeezfan1, Yeagh, Yiwenhuang113, Yoasif, Youatemypotatoes, Yuhong, Zacherik, Zackkattackk247, Zackyboy96, Zalinda Zenobia, Zarmanto, Zebruh, Zekeman95, Zig973, Zinoviev, Zisa, Zlemming, Zm1573, Zoicon5, Zolotros, Zomganairship, Zondor, ZooFari, Zschernicky, Zsinj, Zwynky, Zzuuzz, Ævar Arnfjörð Bjarmason, Œ∑´®†¥πøˆ, ﻢﺣﺑﻭﺑ ﻉﺍﻟﻣ, 2843 anonymous edits

**Audi** *Source*: http://en.wikipedia.org/w/index.php?title=Audi *Contributors*: 02bjahed, 16@r, 1980 Canadian Cup Series, 5 albert square, 72Dino, 7D HMS, 842U, A little insignificant, A to z123, A.K.R., A6guy, ANGELUS, Aa2-2004, Aaroncrick, AbsolutDan, Acroterion, Adambro, Adashiel, AdidasTrainers, AdjustShift, Adrian M. H., Adtroy, Aeusoes1, Agnermk, Ahoerstemeier, Akadruid, Akrabbim, Alai, Alandiggory, Alansohn, Aldis90, Alekseihorn, Alpha77a, Amhainn, Amniarix, Angelice2, Angr, AniRaptor2001, Anna Lincoln, Aptclap25, Arado, Arakunem, Arcfire, Arjayay, Arpingstone, Art LaPella, Arthena, Arthurseabrasil, Articlia, AshveenBansal, Aswinbahudana, Atif.t2, Atril, Attilios, Audi152, AudiS4BiTurbo, Avala, BL, BM99, BSI, Badapro, Badgernet, Bahi, Bear133, Beetstra, Beifahrer, Ben Ben, Bender235, Bernardd55, Big-Pat, Biker Biker, Bilbo571, Billcarr178, Black Kite, Bobblewik, Boivie, Bongwarrior, Bookoflove, Boyakacsha, Brad_, Brossow, Bryan Derksen, Buddha24, Bumper12, Bunnyhop11, Burgundavia, Bwilkins, C3o, CWenger, CWii, CZmarlin, Calvin Winstanley, Capricorn42, Captain Caveman, Captaintomo, Carnity, Carpictures, Carzmaniac, Catgut, Ccunning, Cflm001, Ch Th Jo, Charles01, CharonX, Chav0000, Chelling, Chester Markel, Chowbok, Chris 73, Chris Capoccia, Chris the speller, ChrisGualtieri, Christian Giersing, Christian75, Christienne54, Christopher140691, Christopherdmartin, Chrysalis, Chryslerforever1988, Chuggsymalone, Chuq, Ciphers, Ckatz, Clark89, Clipmode, Colinstu, Colonies Chris, Commodore Sloat, CommonsDelinker, Connormah, Conservative S & R, Conversion script, Cool Blue, Corey Salzano, Cournazac, Crap-parking, Cryogen, Cuski, D, DB 103245, DMG413, Daniel-crawford, Darz Mol, Dave Runger, Dawnseeker2000, DeLarge, Dekisugi, Delirium, Denis1210, Denisarona, Deor, DerHerrLehmann, Deville, Diberri, Diego440, Discospinster, Dj67308, DocWatson42, Domcookmcfc, Donner, Doradus, Dotmarcel, DoxTxob, Dpg10, Dr.C.Winstanley, DrFrench, DragonHawk, Drdisque, Dreamweaverjack, Dreish, Drmies, Dwi Secundus, Dyl, ESkog, Elsonlau, EmilyWolff, Emote, Enigmaman, Epolk, Eric-Wester, EricS, Escape Orbit, Euchiasmus, Evananne, Evanrgray, Ezhiki, Factorylad, Falltillllfy, Fastily, Favonian, Fedayee, Felixstep101, Femto, FiLtroS, Fiachra10003, Filipek, Finbarh2, Fkx, Flewis, Flinnja, Folksong, Fox1, Francs2000, Frazzydee, Fred Bradstadt, Fredrik, Fxhomie, Gahar, Gene Nygaard, Gentgeen, Ghimboueils, Gilliam, Gimmetrow, Gjs238, Glane23, Gr1st, GraemeL, Grafen, Gragox, Gratom, Green slash, Greysonaudi, Gurklurk, Gzornenplatz, Hadal, Halibutt, HamburgerRadio, Hamstar, HannaGrout, Hariva, Hdt83, Hede2000, Hektor, Hisaon, Honza Záruba, Hornymanatee, Hs19833, Hsaunderster, Hu12, Huge colin, Hut 8.5, IAnton7, Ianris, Icecoldsteve732, Igoldste, Infinite34, Irishguy, Ivan10222, J.delanoy, JDC808, JForget, JYing, Ja 62, Jack O'Neill, Jacob Poon, Jamcib, Jaranda, Jared Preston, Jasjit dhillon, Jcarr99, Jcp20, Jcs, Jeff G., Jepku, Jim1138, Jivecat, Jmcc150, Jnc, Joe C, Johann Wolfgang, John, John Nevard, Johnie, Johnuniq, Joseph Solis in Australia, Joshbaumgartner, Jouster, Jovianeye, Jst15, Judae1, KAMiKAZOW, Kaloyantodorov, Kanags, Kariteh, Kayman16, Kbdank71, Keras, Kevin, Kierant, KinsmanRedeemer, Kiwi128, Klemen Kocjancic, Koavf, Koman90, Kozuch, Kristen Eriksen, Krukouski, Kshitij85, Kukini, Kuru, Kwamikagami, L Kensington, LACP LLC, LMB, Lambton, Lavenderbunny, Law, Lawandeconomics1, Laxplayyer23, LeaveSleaves, LeeZ, Letdorf, Leventozler, Lfh, Lhoaxt, Libertycents, Liftam, LighthouseJ, Lightmouse, Liltigger88, Ling.Nut, Lisadezzy, LittleOldMe, Lkesteloot, Lord Emsworth, Lordprice, Lotje, LuxNevada, MER-C, MLRoach, MTan355, MXL52, Mac, Madman212, Madmonky, Magioladitis, Magister Mathematicae, Magnum2037, Magnus.de, Mahdipur, Makwy2, MalcolmStrachan, Malcolma, Malo, Mapleburner21, Marc Lacoste, Marco Krohn, Mark in wiki, Mark83, Markkawika, Mashing pumpkins, Matt Gies, MattWright, Matthead, Mattmils03, Maubertiii, Mav, McSly, Mendaliv, Mengjusha, Mephistophelian, Mfb0ss, Michael Frind, Michelet, Mikaey, Milkmandan, Millisits, Miniotx, Minna Sora no Shita, Mirv, Mitch Ames, Mjkubba, Mjl0509, Mjwalshe, Mkhbgpa, Modster, Monedula, MonkeyKingBar, MonkeyMumford, Monzadave, Moondyne, Moribunt, Morio, Morven, Motacilla, Mr.choppers, MrDolomite, MrOllie, MrRadioGuy, Mrceleb2007, Murtaz lux, Mushin, Mwanner, Mxcatania, N328KF, NPrice, NaBUru38, NativeForeigner, NawlinWiki, Neilwightman, Nephron, Nerval, NetguruDD, Ng.j, Nick Number, Night Gyr, Nikai, Nilesh007r, Nimbusania, Niyathy, Nlegault, Noah Salzman, NocturnalA6 2.7, Nono64, Norm, Nukeless, OSX, Odames866, Odii, Ohconfucius, Ohnoitsjamie, Olivier, Op47, Oroso, Oxymoron83, Palapa, Palnu, Panther 111, Parkjunwung, Parts4euro.com, Patar knight, Paul Richter, Pauli133, Pb30, Pc13, Pechspilz, Penhollow, Pgan002, Piano non troppo, Pikador3C, Pinethicket, Pjedicke, Pontificateus, Porsche997SBS, Prabu Muthusamy, Princip1, Prolog, Pufferling, Pusserstaefook, Quadell, Quadratic, R'n'B, RBNORTIN, RUL3R, RX7000, RadicalBender, Rajakarier, Rammyjames, Ramona ch, Randrewfowler, RapidAssistant, Rapido, RaveDog, RaymondSutanto, RedWolf, Redd Dragon, ReinforcedReinforcements, Related News, RememberCharlie, Remnar, Res2216firestar, Retiono Virginian, Retired user 0002, Reutiadr, Rewster, RexNL, Rich Farmbrough, Rjwilmsi, Rklawton, Rkt2312, Roadrunner, Rocket000, Rodders1991, Romilstr29, RttlesnkeWhiskey, S, SDC, SHCarter, Salome5764, Salondemaria, Sam Hocevar, SamH, Samoborac, Santacruzsurf, Sardanaphalus, Sargentsays, Saucepan, SaviorSix, SchfiftyThree, SchreiberBike, Scieberking, ScottyBerg, Scoty6776, Scriberius, Seann, Seaphoto, Sebastian scha., Setanta747, Sfoskett, Sguler, ShakingSpirit, Shashank751, Shinju, Shirik, Shizane, Shriram, Sibbsy, SidP, Signal11, SimonP, Sjohnson90, Skarebo, Smartboyinuk, Smashville, Smeagol1994, Smilers, Smithy23011996, SoKashira, Socialservice, Solphusion, South Bay, Speer320, Springnuts, Spute, Ssbohio, Stang456, Steffen Kaufmann, Stepho-wrs, Stmauritz, Stombs, Storm Rider, Stormie, StraughanS, Suffusion of Yellow, Surfingslovak, Swaq, Swat671, Swhite1987, Swissman564, SynergyStar, Syrthiss, Szyslak, Taifarious1, Taliska, Tech2blog, Technopat, Teddy.Coughlin, Tejassawant, Teles, Teutonic Tamer, Texture, Tharkibo, The Deviant, The Helper S, The Thing That Should Not Be, The359, TheAutoJunkie, TheBigJagielka, TheK, TheM-SideDon, Thetruthhurts88, Thingg, Tide rolls, TigerShark, Tilly Balent, Tim1357, Timeineurope, Tinton5, Tnxman307, Toddst1, TomS TDotO, Tomh009, Toon nutter, Trendpimp, Tri400, Trident13, Trifecta123, Triwbe, Turk oğlan, Turkforlife, Typ932, Ukn75, UrPQ31, VMG4, VQuakr, Veejay32, Vegaswikian, Verces, Vespristiano, Vibhugupta, Vicoleta, Vijaysub, Vinujadhav, Vivekr13, WHS, Ward3001, Wavelength, Weedle McHairybug, WideArc, WiiKiBoyz, Wikipelli, Wildthing61476, Willirennen, Wilwhymurmoe, Wjs.william, Wmahan, Woohookitty, Wrp103, Wtmitchell, Xboxlive1234, Xtefan, YCCHAN, Yakksoho, Yaris678, Yooden, Youmils03, Yuckfoo, Yuri Gouveia Ribeiro, Yyy, ZapThunderstrike, Zer0431, Zer3k, Zidonuke, ZimZalaBim, ZooFari, Zzuuzz, Σ, Александър, 1344 anonymous edits

**Wrigley_Company** *Source*: http://en.wikipedia.org/w/index.php?title=Wrigley_Company *Contributors*: Acroterion, Ajuk, Amcl, Arite, Ark25, Astatine-210, Asterion, Bavcevic, Beetstra, Bhadani, BigT2006, Bigturtle, Bluemask, Boothy443, Brian the Editor, Bull-Doser, Buzzer2007, CR85747, Cheung1303, Choster, ChrisGualtieri, Connormah, CopperSquare, Craitman17, Dawn Bard, DeadEyeArrow, Diasimon2003, Dr.frog, Elocina, Emperor of Europe, Erianna, Esrever, Evice, Excirial, Frehley, Futurehouse, Gdsamp, Gilliam, Heegoop, Hmains, Hurricanefloyd, Hvn0413, Hydrargyrum, Hydraton31, Ideaslondon, Infrogmation, J Di, JMyrleFuller, JaGa, Jafeluv, Jak86, Jarvishunt, Jasenlee, JazzD, Jerryseinfeld, John, Juliancolton, Kevyn, Kjinho213, Lancer54, Ling.Nut, Luckyz, Lyght, MSTCrow, Man It's So Loud In Here, Matt von Furrie, Matthäus Wander, MetCorp, Mets501, MightyWarrior, Mrceleb2007, NastalgicCam, Nicster09, Nobody of Consequence, NoseNuggets, Nwa757, Oo64eva, Orlady, Pacian, Pcarter7, Piano non troppo, Pig de Wig, Planetstasiak, RadicalBender, Rbpolsen, Realg187, Richieman, Richmond96, Rklawton, RobJ1981, Rockfang, Rossami, SNIyer12, Samuell, Seasee, Secfan, Senator2029, Sherool, Shultzc, SidP, SiobhanHansa, Stephenlau, Stormbay, Svntnth, THEunique, TMV943, Tad Lincoln, Takiman, TechLight, Tennekis, The wub, Tide rolls, Timc, Tinton5, Tkent91, TonyTheTiger, Vippylaman, Wackymacs, War wizard90, Wavelength, WhisperToMe, Wingfatwaila, Working.fool, X201, YUL89YYZ, Yellowdesk, Zannian2000, Zilog Jones, 208 anonymous edits

**Personal_computer_hardware** *Source*: http://en.wikipedia.org/w/index.php?title=Personal_computer_hardware *Contributors*: 1297, 16@r, 1ForTheMoney, 28421u2232nfenfcenc, 2D, 7, ABF, ASmartKid, Aflin, Ahoerstemeier, Aitias, Al E., Alan Liefting, Alansohn, Aldie, Aleenf1, Alfio, Ali, AlistairMcMillan, Almafeta, Alphachimp, AnakngAraw, Anaxial, Andrejj, Andres, Aniket ray, Ansell, Anth.elias, Arda Xi, Aremith, ArnoldReinhold, Aronzak, Arthena, Arunsingh16, Asephei, Ash211, Astragth, Atlant, Atomskninja, Augustojv, Avono, B4hand, Baa, Babajazz!, Barras, Beastfromouteast, Ben-Zin, Benny476, Bennybp, Betacommand, Betodd, Bgutah, Big Brother 1984, Blaxthos, Bluerasberry, Bobbybobington, Bobo192, Bongomatic, Bongwarrior, Brat32, Bsadowski1, Burzmali, Bushcarrot, Buttered Bread, CIreland, Cabbatime, Calabe1992, Calmer Waters, Can't sleep, clown will eat me, Canderson7, CanisRufus, Capricorn42, CaptainVindaloo, CardinalDan, Carinemily, Cdog44, Chiragvyas, Chocolateboy, Chrihern, Chris G, Chzz, Clarince63, ClosedEyesSeeing, Cometstyles, Computerdan000, Computerman45, Conny, Conversion script, Coreycubed, Courcelles, Crazytales, CredoFromStart, Crisco 1492, Cyanoa Crylate, CyberEdZ, D dasun, DARTH SIDIOUS 2, DFRussia, DMacks, Dadude3320, Dalliance, Dantebertana, DarkAudit, Darth Panda, Dasmich82, Daveb, Daveydweeb, David Biddulph, David Gerard, DavidCary, Debug-GED, Deeptrivia, Denisarona, DerHexer, Dicklyon, Dirkbb, Discospinster, DivineAlpha, Dkasak, DocWatson42, Dreadstar, Dskluz, Dysepsion, EJF, ESkog, Edcolins, Edivorce, Edward, Ehheh, Ekashp, El C, Elmindreda, Emhoo, Epbr123, Erik9, EryZ, Escape Orbit, Everbloom, Evlekis, Excirial, Falcon8765, Farosdaughter, Fazdabest, Fedaia, Feydey, Fieldday-sunday, Finalnight, Fluri, Foxjwill, Frap, Fsdfs, Funandtrvl, Fyyer, GB fan, Giftlite, GilbertoSilvaFan, Gilliam, Gimboid13, Glane23, Glass Sword, Glow-in-the-Dark Guy, Golbez, Goldom, Gorgan almighty, Graham87, GrayFullbuster, Groby, Gurch, Gwen Gale, Gwernol, Gökhan, HalfShadow, Happah, Hardware.com, Haritacomputes, Hede2000, HenkeB, Herazy, Heron, HexaChord, Hmains, Hotshot977, Hurting4000, Hydrogen Iodide, I already forgot, I am stuiped, IRP, Ianweller, Icairns, Ilyaroz, Imasleepviking, Immunize, Iridescent, IvanAndreevich, Ixfd64, Izcool, J.delanoy, JLaTondre, Ja 62, Jabencarsey, Jacob2569, Jamesooders, Japonca, Jay, Jeff G., Jennavecia, Jimmi Hugh, Johan1298, John, Johnbrownsbody, Johnman239, JonHarder, Joseph507357, Jpbowen, Jpgordon, Jshi110, Julz888, Jumpman23.1313, Jusdafax, Jwoodger, Jxw13, Kanags, KathrynLybarger, Kbdank71, KeKe, Kenny sh, Kevinb, Kingadam150, Kingpin13, Kirachinmoku, Kizor, Knownot, Knuckx, Knutux, Kozuch, Kromped, Kubanczyk, Kuyamarco123, Kwiki, L Kensington, LaggedOnUser, Landon1980, Lanner10, LeaveSleaves, Leedeth, Leetfiber13, LiDaobing, Likepavan, LiteralKa, Logan, Logicwiki, Luckas Blade, Luke poa, Luna Santin, Lurker, Luís Felipe Braga, M Nabil, M4gnum0n, MANGEJ KUMAR, MER-C, MJ94, MMuzammils, MUdayKiran, Ma8thew, Mac, Madhero88, Maelnuneb, Majorclanger, Makeemlighter, Mammiyydaddiyy, Manishearth, Manjudeviprasad, Mannafredo, MarSch, Marcelo-Silva, Marcschulz, Martin451, MassimoAr, Mato, Matt Britt, MaverickSolutions, Mayur, Mekong Bluesman, Mentifisto, Mhare, Michael Hardy, MichaelMcGuffin, Midnightcomm, Mike Rosoft, Mikeh0303, Miko3k, Miquonranger03, Monkee13, Monkey Bounce, Monobi, Montgomery '39, Mpgenius, Mpradeep, Mr Stephen, Mr.Anderson-Queens-AQ, MrOllie, Mrzaius, Msajouher, Muhandes, Mushroom, Music Sorter, Mustafaa, Mygerardromance, Myscrnnm, Mythealias, Nancycil, Nanshu, NatureA16, NawlinWiki, Nemilar, NerdyScienceDude, Neurolysis, Nicholsr, Nickdella2, Nixdorf, No1lakersfan, Nonforma, Nopetro, Notjake13, Novum, Numbo3, Nurg, Odie5533, Ohnoitsjamie, OisinisiO, Olie93, Oliver202, Oliverdl, Omicronpersei8, Onorem, OpenInfoForAll, OrgasGirl, Oxymoron83, PAK Man, PTSE, Paul A, Pcap, Perkelito, Phantomsteve, Phgao, PhilKnight, Philip Trueman, Piano non troppo, Pinethicket, Pixel8, Pnm, Pookleblinky, Possum, Psantora, Pujar123, Pyrospirit, Qirex, Quokly, Qxz, R, R04drunerr, Rabit gti, Radagast83, Rahul0162, RainbowOfLight, Rajshree Dhapke, RaseaC, Ray Van De Walker, Rchandra, Rcsprinter123, Rebroad, Reconsider the static, Reddi, Rettetast, RexNL, Rhondah2011, Rick Sidwell, Ridge Runner, Rjwilmsi, Rkinder, Rmosler2100, Robert Merkel, Robinh, Robomaeyhem, Rodhullandemu, Rsm99833, Rwwww, SReznikoff, Saddy Dumpington, Sajith...follows, Saketmitra, Sap123, SchreyP, SchuminWeb, Scott McNay, Scott Wilson, Sct72, Seaphoto, Seidenstud, Selkem, Shadygrove2007, Shervj2, Sietse Snel, Silverxxx, Sithod, Skeetch, Skizzik, Sleeping143, Slon02, Smalljim, Smiteri, Snagari, Snigbrook, Softtest123, Sooriyank, Sophus Bie, Sourside21, South Bay, SpaceFlight89, SpikeToronto, Srini.s1994, Staxringold, StephenBuxton,

Stephenb, SuperHamster, Sword, Szajd, TEB728, Tandral, Tatrgel, Tcncv, Techman224, Teehee123, Tgw333, ThaddeusB, Tharcore, The Anome, The Firewall, The Random Editor, The Sanctuary Sparrow, The Thing That Should Not Be, The undertow, TheBurningFlames, TheGhostOfXmasPast, TheInfernoX, Thecheesykid, Therag18, Thingg, Thr4wn, Thunderboltz, Tiddly Tom, Tide rolls, TimTIm, Tins128, Tirlias, Tonyiswac, Traroth, Treisijs, Ukexpat, UnfriendlyFire, Useight, Utcursch, Vald, Vkurtz, Volker, WIN2012-CONT490-001, WLU, WOSlinker, Wafulz, WaitingForConnection, Washburnmav, Wavelength, Wbm1058, Weregerbil, Wernher, Western Pines, Where, Whisky drinker, WikiJaZon, Wikipelli, Wikitanvir, Wikitumnus, Wimt, WinampLlama, Wjejskenewr, Wknight94, WojPob, Woland37, Woohookitty, WookieInHeat, Wtshymanski, Wyco, Xe7al, Xxray03, Xyzzyplugh, Yakudza, Yesukai, Yesval, Ylem, Zak1551, Zfr, Zhou Yu, Ziusudra, Zlajas, Zollerriia, Zyro70, Zzuuzz, ديسلا ىفطصم دمحا, 1345 anonymous edits

**Dock_Connector** *Source*: http://en.wikipedia.org/w/index.php?title=Dock_Connector *Contributors*: Ajfweb, AvicAWB, Boock, Chzz, Dougmcdonell, Drmies, Electronet1, Epson291, Eraserhead1, Evosoho, Fourthords, ISquishy, InternetMeme, J Clear, John W. Kennedy, Marcus Qwertyus, Mastermind3895, Mdwh, Miami33139, Minnaert, Myscrnnm, PatrickJCollins, Sir Stig, SteveCostello, Tamer of hope, Ua hockey, WWriter, Whywhenwhohow, 33 anonymous edits

**IPod Classic** *Source*: http://en.wikipedia.org/w/index.php?title=IPod_Classic *Contributors*: .telus., 808-Thug, A. Exeunt, Acps110, Addihockey10, Aido2002, Airplaneman, Alansohn, AlistairMcMillan, Allibyrd, Amcbride, Amorymeltzer, Anarchy 97, Anetode, AngelOfSadness, Angrysockhop, Anna Lincoln, Aofjsye, AquaStreak, Arthena, Asav, Ashleighp, Avb, Baa, Bacchus87, Bairdjb, Bajaguy, Banalnatae, Baseball Bugs, Bbrriiaannhhaammmm, Bdamokos, Bencey, Bidofthis, Bilou, Bloodbroad, Bobianite, Boomtown Rat, Bramo126, BuickCenturyDriver, Butterfly0fdoom, CL, Camelcast, Cerebrawl, Chnt, Chris as I am Chris, Ckatz, Clarince63, Clayoquot, Coffee, Collideasope, CommonsDelinker, Consumed Crustacean, Crazy4iPod, Csaag, Cycomachead, Danhash, Danity94, DavidRF, Dazzafar, DeadEyeArrow, Dekisugi, Devils471, Dgies, Djdjcjcj3421, Djmckee1, Dominicfitz, Douglasr007, Drama123, Drmies, Dudesleeper, E Wing, East718, Echuck215, Eeekster, Eehenry2, Eeshkol, Elliskev, Emason16, Epbr123, Epson291, Eraserhead1, Eric-Wester, Esanchez7587, Escape Orbit, EternalRains, Ethanpilot787, Evosoho, Excirial, Eyesbomb, Fatnegrolips, Ffiti, Firekite, FisherQueen, Flewis, Fluffernutter, Foetusized, Fritzpoll, Fury, Fvasconcellos, GB fan, Gbrading, Geek45, Geoking66, Glane23, Glennforever, Gogo Dodo, Grafen, Groink, Guybrush, HX368, Hayden120, Hbdragon88, Headbomb, Henrikb4, HereToHelp, Hirak 99, Hojx93, Horserice, Hroyer, Hut 8.5, I am pieman, IGeMiNix, ILoza, Icanedit2, Illegal Operation, Imaginists, In23065, Indon, Infuriated-Z, Ipodlord, Iridescent, It Is Me Here, Iuhkjhk87y678, J.delanoy, J4musicals, JNW, JOSamsung, JP Godfrey, JQF, JRC3, James Kanjo, Jameswa21, Jamiesterling, Jay794, Jdtyler, Jeff G., Jherm, Jhsounds, Jimthing, Jj137, Jksun2, John Quiggin, JohnnyBGood, Joker217, Jondo220, Jstupple7, Jtbean, Julian BH, Jw21, Kafka212121, Kanesue, KeineLust90, KelleyCook, Kendal Ozzel, Kesac, Kevin S., Kingchip101, Kitch, Klubclub1, Kungming2, Kyro, L Kensington, L337p4wn, LFiveZeroFive, LOLDSFAN, La Pianista, Lawrence Davis, Leathwick, LethalReflex, Lightmouse, Lil Lez, Ll0l00l, Logan, Lt. penguin, Luk, LukeShu, Lynk14, M1ss1ontomars2k4, MJisnotmylover, MPerel, Ma8thew, Macaddct1984, MacsBug, Mandarax, MantaRay, Marcus Brute, Materialscientist, Matthew98, Max Naylor, Mduser63, Megamanfan3, Melesse, Melmann, Memphisto, Mentalman1369, Mhking, Michaelscott333, Mike.lifeguard, Mild Bill Hiccup, Minna Sora no Shita, Minnaarpieters, Miquonranger03, Mister macphisto, Monitorer, Mono, Moozaad, Moronicles, Mugunth Kumar, Mynameisalex01, Mynameisorange, Myscrnnm, Mysdaao, N i grunze, NCurse, NHJG, Naruto Tron, NawlinWiki, NerdyScienceDude, NewYorkDreams, Niedson, Nklapproth, Not Chrissy, NyyDave, Ohconfucius, Ohnoitsjamie, Oiyou17, Omar6000, Oscarthecat, Oxymoron83, PGSONIC, Pachelbel224, Pacific Coast Highway, Pamela Gardiner, Papre, Pele Merengue, Peter Karlsen, Pharos, Phoenix2, Photoguy439, Pinkadelica, PlazzTT, Plrk, ProductResource, Psantora, Pseudomonas, Rapido, Rawr, Regipalka, Rehcsif, Rehevkor, Res2216firestar, Richi, Rigadoun, Ripe, Rje, RobWill80, Rokfaith, Rollo Bay 1758, Ronhjones, Root212, Rottwieler16, Rrburke, Rronalds, Ryanonfire, SFfreak, Safesler, Saksi88, Samsoncity, SchfiftyThree, ScottJ, Sefrey, Seraphim, Shazam619, Shieber, Shiori, Skunkboy74, Sniper Zed, Snoilog, SoWhy, Sound of white, Spartan959, Spcleddy, Speer320, SpikeToronto, Staffwaterboy, Stefan Kruithof, Stephen Shaw, SteveSims, Stevenrasnick, Stick9Figure, Stickguy, Storm1879, Stormx2, Sween64, Swift as an Eagle, T.J.V., Tamajared, Tbgkol, Tbhotch, Techman224, The Thing That Should Not Be, The Wild Falcon, The iPod Color, TheHYPO, TheHerbalGerbil, TheSeer, Thekmc, Thekoyaanisqatsi, Theunamedguy, Thumperward, Thunderbird2, Tide rolls, Tingrin87, Tokek, TonyTheTiger, Tpbradbury, Trevj, Typhoonchaser, Undc23, Unr3a1r00t, Untitledmind72, Urhixidur, Vanished User 1004, Velella, VideoCWK, ViperSnake151, Wannger27, Wellesradio, West.andrew.g, Whenbongoscollide, WikiDude121, Willirennen, Xdenizen, Xeno, Xxhopingtearsxx, Yacoob, Yellowdesk, Zachzintel, Zippanova, Zodon, Zxcvboy97, 783 anonymous edits

**IPod Touch** *Source*: http://en.wikipedia.org/w/index.php?title=IPod_Touch *Contributors*: $01734071290912$, *Kat*, -a-w-13, .telus., 10metreh, 15avaughn, 1diesel, 1exec1, 28421u2232nfenfcenc, 333cool, 382mayanzev, 5 albert square, 7OA, 808-Thug, 918577, 9Nak, A5b, A7x, A8UDI, A930913, AContributor, AMK1211, ARC Gritt, ASA-IRULE, ASOTMKX, Aaa321aaa, Aalegado, Aaronrp, Abb615, Abu-Dun, Acalamari, Accounting4Taste, Acps110, Acroterion, Adammcg, Addihockey10, Addit, Aeternus, Aff123a, Ageekgal, Aglyad, Ahmad.j.ziq, Ahomentc, Aido2002, Aitias, Aizuku, Ajayfahlman, Ajfweb, Aka042, Alan Liefting, Alansohn, Albany NY, Ale jrb, Alecao89, Aleenf1, Alexafhalex, Alexbuirds, Alexishot, Alfstar1997, AlistairMcMillan, Allens, Allmightyduck, Allstarecho, Alphathon, Altenhofen, Altintx, Altzinn, Alēex17, Amdrummer80, Ameliorate!, Amog, Anarchy 97, Anaxial, Andreasravndal, Andrei S, Anetode, Animal91X, Anna Lincoln, Anonymous Dissident, Anpecota, Antonov86, Apena16, Aperson1234567, AppleSuckz, Appleuser4, Aprock, Aquafiredragon, Aquillyne, Aquilosion, Aradius, ArchonMagnus, Arekfal, Aremith, Ark25, ArmyOfPie, Art LaPella, Artatac, Artemka373, ArtistScientist, Aruhnka, Asderoff, Asdfadsfadsfadsf, Ashill, Ashraf Hakim, Astonmartini, Astroview120mm, AustinZ, Ausuomi, Autoerrant, Avazelda13, Avicennasis, Awesomeness95, AxG, Azimuth1, B, B00bz1995, B4lyphe, BAM9412, BK DC, Ballin20914, Ballonlevi69, Bamrocks01, Banjaloupe, Banksbr2, Barek, BarryTheUnicorn, Basawala, Baseball-bob, Baseballboy23ak, Bassadams, Bassbonerocks, Bazzargh, Bbb2007, Beao, Beautiful so ur, Ben Ben, Ben420, Bencey, Bender235, Benjaminevans82, Benstown, Bentoman, Betacommand, Bgold, Bigb0i318, Bigedak, Bigghuggepennissfun69ers, Bihco, Bingobangobongoboo, Birdy541, BlaiseFEgan, BlakeKarter, Blakely07633, BlastOButter42, Blb9556, Bliks, Bluedeathcholera, Bluezy, Bmusician, Bmz loop, Bobianite, Bobo192, Bobpic8, Boccobrock, Bogidu, Bongwarrior, Borhos222, Boxstaa, Brain seltzer, Brandmeister, Breawycker, Brereton55, BrianCMasi, BrianRecchia, Brianreading, Briguychau, Britannic124, Brittflick 1994, BrokenSphere, Brookie, Broonie, Brusegadi, Bryantsinger, Budsully13, Bugefun, BullRangifer, Burnte, Butterfly0fdoom, CALQL8, CIreland, CRobClark, CTho, CWenger, CZMQFRG, Cabehead12, Caiaffa, Calabe1992, Calador109, CamCreator, Camanda, Cameron Scott, Cameroni30, Canterbury Tail, Canukk14, Capricorn42, Captain panda, Captainrag, Casamort, CaseyPenk, Caspertheghost, Catgut, Celzey12, Cflm001, Chambo622, ChaosSorcerer91, ChaoticGhost, Charlielowry, Chew Vader, Chilliwilly1000, Chillllls, Chpfeiffer, Chris Price, Chris as I am Chris, Chris-Gonzales, Chris01720, Chriscombs, Christ904, Christian75, Chzz, CityOfSilver, Cjo123, Ckatz, Clay 767, Closedmouth, Clovis Sangrail, Clsmooth2, Cobraman156, Coldpower27, Colejohnson66, Colenso, ColoredCrayon, Commander Shepard, CommonsDelinker, Companioncube31, Conscious, CoolDude330, Cooperjct, Cpwhaward, Crab135, CrazyChemGuy, Crazytales, Crazyviolinist, Cremepuff222, Crisss1104, Crobichaud, Csaag, Csmith1994, Csorfab, Cstaffa, Cuba46, CurlyGirl93, Cwicestory, Cybercobra, Cyberius, Cyrus XIII, DAJF, DARTH SIDIOUS 2, DVdm, Dacrone, Dadioles, Dajavax, Dakota81, Dale Arnett, Dalton358, Dalton4554, Damian Yerrick, Damicatz, Dancter, Daniel J. Leivick, DanielDPeterson, Darijoe, Dario D., DarkAudit, Darkwind, Darkxsun, Darry2385, Darrylpoh, Darth Metus, Darth Panda, David, David Schaich, Davidpar, Davylu777, Dawn Bard, Dawnseeker2000, Dcederberg, Deandbest, Death8689, DeathToEnemies, Debvgger, Dekisugi, Delcity, Derafols, Destin, Destynova, Deutschgirl, Df53515, Dgies, Dhscommtech, Diamondblade2008, Diannaa, Dingting10, Dinkytown, Discospinster, DisneyCSIfan, DividedByNegativeZero, DivineAlpha, Djdannyp, Djones96, Dlae, Dmhsfan, DogGunn, Dontgetownt, Dopplerd, Doublex120, Douglasrmcdonald, DougsTech, Dr.K., Dragana666, DragonflySixtyseven, Dragonflydragon, Dream out loud, Dreamyshade, Drewson99, Drilnoth, Drmies, Droobey, Drpepper100, Drrll, Drumpat01, Dsdukcy, Dspradau, Duckman025, Ducleus, Dude963841, Duncan, Dustin Howett, Dvdhn, E. Ripley, E0steven, EAi, EVula, Eb.eric, Edderso, Editoer, Editor182, Edsthing2, Eggman123, Ehdr, Ehsteve23, Eibcga, Ekevu, ElKevbo, Elcstone, Elektron, Elkan viel, Elvey, Elwood00, Eman297, EnglishDude, Enviroboy, EoGuy, EoffreyGey, Epbr123, Epson291, EqualRights, Equazcion, Eraserhead1, Erianna, Eric 7666, Eric-Wester, Erifneerg, Errorlines, Erutuon, Esanchez7587, Escape Orbit, Ethanpilot787, Evanherk, Evannxmcrr, Everlasting enigma, Exar kaun, Exavier24, Excirial, Exciter1001, Explicit, F1racer101, FCYTravis, FF2010, Falcon8765, Falcon9x5, Fasach Nua, FastKarts, Fastily, FatalError, Favonian, Fenrir-of-the-Shadows, Fenwayguy, FeralDruid, Ferthewin, Fetchcomms, Fefour, Ffrogman, Fieldday-sunday, Fighting for Justice, Filing Flunky, Fire 55, Firebringervt, Fish6334, Fitness = Health, Fletcher, Fluteflute, Flyguyty89, Flyinhawaiian, Fnagaton, Forallilove1, Forteblast, Fosterec, Frankenpuppy, Fredclaus1234567890, Free dude, Freegon427, FreewareLovers, Frehley, Freshpop, Frigotoni, FromFoamsToWaves, Frozen-Coke-Rocks, Fruitking98, Fullyrandomtandem, Fumitol, Fumo7887, Future Perfect at Sunrise, Fyyer, GB fan, GDallimore, GRBerry, GRevolution824, Gail, GainLine, Gamefreak2413, Gamersedge, Garrett247, Gary King, Gators2321, Gaul, Gbelk08, Genius101, Gggh, Gh5046, Gilliam, Gio1766, Giogiogio4, Girlfriend Thief, Gm206, GnomesRuleTheEarth, Goffrie, Gogglemaster, Gogo Dodo, GoingBatty, Gokupiccilo, Golfertig, Goosnarrggh, Grafen, GrapeSteinbeck, Greg Salter, Groink, Grue.z, Gscshoyru, Gsp8181, Gtr9, Gtstricky, Guat6, Guess Who, Guinea pig warrior, Gujamin, Gunmetal Angel, Gurch, Guy Harris, Guyinchair, Gwen Gale, Gwernol, Gyrferret, Gökhan, H92, HFret, HJ Mitchell, Hagerman, HalfShadow, Hallows AG, Hamish and andy, HangingCurve, HansTAR, Hapmt, Hardyplants, Harmelion, Haseo9999, Hateless, Hatrox, Hauzer-Ninja, Hbent, Hcashmore11, HedgehogHK, Heirpixel, HellDragon, Hellish rabbit, Henry W. Schmitt, HereToHelp, HexaChord, HiDrNick, Hiddenfromview, Higgins1095, Himan123856, Hippo99, Hittingray, Hmrox, Hoary, Hoho832, Hongkongeditor, Hornnic, Horserice, Hrajt2, Huangcjz, Hunteris2dope, Hut 8.5, II MusLiM HyBRiD II, ILoza, IOSFreedom, IRP, ISquishy, ITurtle, IW.HG, Ibamberger, IbrahimElsherbiny, Icarbohydrate, Ieyetroll, Ignorance is strength, Iloveaquagreen, Ilyanep, In23065, Indy500noah, Infomantoday, Infosprt, Inspector 34, InternationalLog, Inuboy1000, Ipodsucks, Ipodtouch4 blah, Ipopod, Iptouch4, Irfansweb, Iridescent, Irishguy, Ironist, Islander, Itaiyz97, Ivucica, Ixfd64, Izno, Izzi303, Izzixe, J.delanoy, J.p.omalley1996, J4m35 80nd, JElvidge, JForget, JGXenite, JLaTondre, Jabberwock Jack, Jackfork, JackieChiles7, Jackster, Jake Wartenberg, Jakedemian, Jakeiscool86, Jakerdemian, Jakestedman, James086, JamesBWatson, Jamil d, Jarisony, Jarrodpi, Jason Garrick, JasonAQuest, Javert, Jawadreventon, Jaydentaku, Jchubby, Jeff G., Jeffpower, Jeffwang, Jellis131, Jeremy Visser, Jerry teps, Jerryobject, Jesuscampbell, JetLover, Jezbcc, Jfv93, Jgcena, Jhsounds, Jim10701, Jimscott09, Jimscott13, Jimthing, Jisp, Jk2q3jrklse, Jlc0023, Jlocalled, Jmcollier, Jmicdizz, Jmphillips99, Jmrowland, Jncraton, Jncro, Joeclark, Joeemslie, John John 216, John W. Kennedy, John254, John72rio, Johnbrap, Johndark93, Johngibbs2010, Johnsu01, Johnyrock17, Joinuseveryweek, Jojhutton, Jonasftw, Jonjesbuzz, Jonnyboy88, Jonnycohen, Jons63, Jose Concepcion, Josemanimala, Joshschr, Joshua1900, Joshygoodwin, Joshyworld, Jossi, Jotomicron, Jpppwiki, Jrdaigle1000, Jreferee, Jrtayloriv, Jslefo, Jsnell, Jtaraszka, Jtbean, Julien Deveraux, Jusdafax, Justinxtreme, Jw21, KDesk, KGasso, KJRehberg, Kaiba, Kalmenius, Kanonkas, Kartiksingh1992, Katoh, Katori22, Kav-de-moose, Kbandy, KelleyCook, Kendrex, Kevin chen2003, Kilowattradio, King kolton9, Kingpin13, Kjinho213, Kle0012, Klopek007, Kmsiever, KnowledgeOfSelf, Koavf, Kockgunner, Koshoes, Krashlandon, Krazycev13, Ks0stm, Kubigula, KudulO, Kukini, Kurdell, Kuru, Kwamikagami, Kyro, L Kensington, L337p4wn, L33tbobsledder, LAX, LBJtheking, LOctopus, La-UK-Dude, Lamadude, LaptopGun, Larsbaasz, Lasse Havelund, Lazulilasher, Lemmayoshi, Leocheng, Leon7, Leroyinc, Lexischemen, Lhasapso, Lifeofadaydreamer, LilMane, Lilcaroline96, Lineface, Liuofficial, Ljsdcdev, Lodevermeiren, Logan, Lola Voss, Loopy2424, Lord Kelvin, Lostgame, Lowseylos, Lox, Ludo716, Lulu of the Lotus-Eaters, Lxsys, Lynkynpark86, M gol, M.O.X, M0thr4, MC Dupree, MHB210, Ma8thew, Mabilal, Mac Starkiller, Mac128, Macaddct1984, Macedoniarulez, Mackatacka123, Macman16a, Macy, Maddiekate, MadnessInside, Mahanga, Makio88, Malcolmxl5, Malinaccier, Man It's So Loud In Here, Manop, MantaRay, Marcellus2070, Marco Guzman, Jr, Marcus Qwertyus, Marcusliou, Marek69, Mareklug, Markintellect, Markthemac, Marky540, Maroonedsorrow, Marskuzz, Martin451, Marwikitin, Marx01, Masky, Master of Puppets, Masterlockstfu12, Mathiastck, Mathwizxp, Mato, Matt.T, Mattf19, Matthew Yeager, Matthew98, Matthewmiller1234, Mattwright314, Matwjen, Max Naylor, Max.goedjen, Mazlrel, McSly, Mcalexbag, Mcalhanytrey, Mcoupal, Medudeisay, Megamanfan3, Melab-1, Mentifisto, Mephistophelian, MetaDark, Mets501, Metthe, Mgersten, Mghostsoft, Mhoenig, Michaeljoeseph, MickelsonBJ, Miffy900, MightyWarrior, Mikepanhu, Minimac, Minimac93, MinimanDragon32, Miraculousrandomness, Misterbossplaya, Mistercp, Mjbauer95, Mjdf hgjklsdrgaser, Mjf042, Mjtanton, Mks-mk, Mnl831, Mobile Snail, Mod.torrentrealm, MondayLad, Money man1123456789, MonkeyHouse98, Mono, Moocowsrock, Moomoomoo, Moronicles, Mortense, Motomotom, Moviefan222, Mr pig face, Mr.Kennedy1, Mr.Z-man, Mr.bombo, Mrblack927, Mrdonnelly, Mrmoocow44, Mrngoitall, Ms 2007 Cn, MuZemike, Mugunth Kumar, Muhammad baitul alim,

Multi-modz, Mushroom, Mutchy126, Mx3, My Ubuntu, Mygerardromance, Mykel2727, Myscrnnm, N i grunze, N94812, NFreak007, NYMFan69-86, Naggyman, Nanami Kamimura, Nasnema, Navy Blue, NawlinWiki, Ncmvocalist, Nebabc11, Neilt6, NellieBly, Nelsong14, NeoRicen, Neokmc, Nepenthes, NerdyScienceDude, Netknowle, Neverquick, Newguineafan, Newtown11, NexusCFX, Nich148 9, NicholasW1, NickBush24, NickCap, NickW557, Nihiltres, Ninjapunk21, Nitromaster, Nitsram, Nnemo, Nnickn, Noah Salzman, Noahman5120, Noctibus, Nsaa, Nutiketaiel, Nuttycoconut, Ny156uk, Nyeboy9, Oahiyeel, Oback, Obsoletepower, Oddball163, Off we go, Ogigacz, Ohconfucius, Ohnoitsjamie, Olavid, Olli, Ollie31770, Oo nrb, Ophelia88, Opmadaration, OprahDust, Opspin, Optakeover, Orangemike, Orayzio, Orthoepy, Oscarthecat, Oswald07, Otsel7, Ouimetnick, Ourai, Ouro, OverlordQ, Oversizedguitar, Oxymoron83, Ozguy89, PAntoni, PM800, Pah799, PaleAqua, Paranormal Skeptic, Parrotheadmjb, Patrick, Patrickloggins, Patrickrm, Patriots2212, Pattyd1, Pauerbach08, Paulj032, Pavel.nps, PdDemeter, PedanticHipster, PenComputingPerson, Penz22, Persian Poet Gal, Personfromthatplace, Peter92542007, Phearson, Philbury, Phildopos, Philip Trueman, Phonicsmonkey3, Photographerguy, Phuzion, Pi3832, Piano non troppo, PicklePower, Pigsonthewing, Pikajew, Pingveno, PitFan75, Plainchips, Pmsyyz, Pnightingale, Poison the Well, Pol430, PopePmp25, Portia327, Pranaysanghavi, Pratyay, Pretzels, Prissi, Prodego, ProductResource, Proflame, Progressive505, Prophetmike, PsY.cHo, Psantora, PseudoKirby, PseudoNym, PseudoOne, Psyc, Pullshapes, Pumapayam, Pure mayhemica, Pureeminences, Pwntbywombat, Pyro3d, Pyro721, Pyromaniaque, QUILzhunter931, Quadratus, Quags92, QuickPWN, Quinten2741, R'n'B, RFerreira, RHSB Scipio, RJaguar3, Raccettura, Racedude, RachC, RadioFan, Radon210, Ral725, RandomKid39, Randomfrenchie, Randomtime, Rat2, Raveonpraghga, Rawr-e, RazerM, Raztus, Rboyce, Reaper Eternal, Reddevil0728, Redey o, Redgecko20, Redsox74, Rees11, Refmaster, Refsworldlee, Regibox, Rehevkor, Reinthal, Remstar, RepublicanJacobite, Res2216firestar, Resplendent, Reswobslc, RexNL, ReyBrujo, Rhopkins8, Ric36, Ridge Runner, Riffic, Rjd0060, Rjwilmsi, RkOrton, Rmsuperstar99, Robbieg123, Robertvan1, Rockfang, Rockysmile11, Rodhullandemu, Rohan Jayasekera, Romanian Relish, Ronark, Rrburke, Rror, Rtyq2, RubenSchade, Rubiscous, Ruby.red.roses, Rufus843, Rukaribe, Rusted-angel, Rx5674, Ryulong, SD5, SJobs, SOADFan17, ST47, Sam112294, Sambolini, Samkass, Samlikeswiki, Samueltruman, SandBoxTheVideoGame, Sapienyia, SarekOfVulcan, Sarkirk278, Satori Son, Save-Me-Oprah, SaveThePoint, Sawblade767, SaxofoneDL, Sc9999, SchfiftyThree, Scid, Scootey, ScottSteiner, Scottmacpherson, ScottyV89, Sculpher, Sdfisher, Seanrobinson747, Seaphoto, Secret Saturdays, Sennen goroshi, Sevenationarmy17, Sgoh3781, Shadowrex, Shadowriver, Shanafme, Sharkbait491, Shawnc, Shencypeter, Shinedrc, Shinjiman, Shortyeu, Sifaka, Silverturtle1, Simon.hess, Singingaccordionist, Sionus, Sir Stig, Sirdeath139, Sirus20x6, Sjö, Sk8rSoda, Skittle, SkyMachine, SkyWalker, Skyezx, Slakr, SleweD, SlimShadier, Slimey01, Slockpatel, Sman789, SmileyBarry, SmileyDude, Snail Doom, Snorfy, Snowball2979, Snowolf, So God created Manchester, Soboredlesso, Solipse, Some jerk on the Internet, SomeUsr, Someguy432, Sonicblue4, Sonixrulerz, Soojus, Southern Illinois SKYWARN, SovietCanuck, Sp, Spartacus106, Spazturtle, Speakus, Spec ops commando, Speer320, Spesh531, Spidey665, Spitfire, Staffwaterboy, StarWarsDude2007, Stealth500, Steph1010, Stephen Shaw, Steven Walling, Stew13, StewieX, Stickguy, StickyCheese2000, Stocklera1, Storm Rider, Stupid Corn, Stwalkerster, Supa596, Super Shy Guy Bros., Supercool94, SwisterTwister, Szeltman, THEN WHO WAS PHONE?, TJ Spyke, TMV943, TTS51207, Tabledhote, Taintedyouth, Talu42, Tamajared, Tariqabjotu, TaskSpan, Taychunmin, Tbhotch, Tcncv, Tech30, Techman224, Teddy10, Tempodivalse, Tenbergen, Tengilorg, Teni Tadevosyan, Terence1122, Terrillja, TestUser11223344, Tfmurphhk, Tgeairn, Thatappleguy, Thatstheideabob, The Thing That Should Not Be, The Wild Falcon, The cactar, The wub, TheFreeloader, TheNewPhobia, TheRico152, TheSeer, TheTechFan, TheXenocide, Theapplenerd, Thedjbogus, Thefatmould, Thegoldbar, Thehelpfulone, Theklavern, Theleftorium, Thingg, ThirthtonThithtertinton, ThomasHarte, Thomaselite13, Thumperward, Tian2992, Tiddly Tom, Tide rolls, Timmyishappy, Timsdad, Tiptoety, Tkasmai, Tm1000, Todd.Isler, Tofutwitch11, Tohd8BohaithuGh1, Tokek, Tom Morris, Tom743, Tommey22, Tommy2010, Tonker83, Tony Myers, Toon05, ToshibaG450, Toskweeze, Touching is good, Toussaint, Transisto, Trbmagic, Trekky0623, TrevorLSciAct, Trusilver, Tucker001, Tumicking, Turd man to the rescue, TutterMouse, Tweenk, Twexler, Typhoonchaser, U23drox, Ucucha, Uker, Ukexpat, Uncle Milty, Unkwown12, Unseenpunk22, Urbndude, UrboyJDH, Urhixidur, Us4member3, Uspn, Utkarsh apple, V12vanquish, Vampires drink blood, Van helsing, VanishedUser30023402384023, Vchimpanzee, Velella, Versageek, Versus22, Vincent Pun, ViperSnake151, Vishnava, VisvambaNathan, Vneiomazza, Vrenator, WWB, Wackymacs, WadeSimMiser, Waggers, Wagig, Waisybabu, Walkingonthesun, Walter Görlitz, Wamke4, Wannger27, Warut, Wasell, Wayne Slam, We hope, Weh0068, Wernjump, Whatsurstory, White hawk rz89, Whitepaw, Whyzeee, WichitaQ, Wiidestiny, WikiLaurent, Wikievil666, Wikipelli, Wikiuser0123, Wilhelmp, Willking1979, Wj32, Wjejskenewr, Wknight94, Wolfnix, Wordsbywiki, Work permit, WriterHound, Wrong, WryVendor, Wtmitchell, Wysprgr2005, XX55XX, XXXYouXXX, XXmatt69Xx1, Xamphear, Xantosuex, Xatpod, Xeno, Xhead12, XjakusX, Xmoogle, Xnemesis, Xp54321, Yankeesownyou1, Yavrey, Yellowdesk, YesIAmAnIdiot, Yggdriedi, Ylwsub68, Yoahann, Yohowithrum, Youbeat2008, Your Disco Cats, Yugimaster81, ZV1998, Zach Vega, Zackaback, Zakawer, Zanndostaro, Zarcadia, Zeromaru, Ziansh, Zippy120, Zipstar994, Zombiyaic, Zouavman Le Zouave, Zrulli, Zulkaroth, Zxb, Zxcvboy97, Zywxn, Σ, 2986 anonymous edits

**iPod Nano** *Source*: http://en.wikipedia.org/w/index.php?title=IPod_Nano *Contributors*: -Majestic-, 10014derek, 3x0, 78.26, 808-Thug, A Softer Answer, Aaron Brenneman, Aberglaube, Aceospades1250, Acmilan10italia, Aconcagua, Acps110, Adammcg, Addshore, Adrian.benko, AgentCDE, Ahess247, Aido2002, Airplaneman, Aitias, Ajcfreak, Alansohn, Aleenf1, Alegoo92, Alexandria, Alexcooldude, Alfredo J. Herrera Lago, AlistairMcMillan, Alpha Quadrant (alt), Alphax, Alsandro, AmiDaniel, Amorymeltzer, Anc726, Andonic, Andreasravndal, AndrewHowse, AndrewRH, Andy M. Wang, Andyroo161, Anecdote, Anetode, Animum, Anna Lincoln, Anonymous6494, AnotherWaldo, Anymediaguy, Anypodetos, Aofjsye, Aoi, ApolloBoy, Apparition11, AppleMan1, ArielGold, Arjunm, Asderoff, Asenine, Ashertg, Asofdjhkljashdl, Aspeicher17, Astonmartini, AtheWeatherman, Athletechat, AutoAstro1981, Awesomeness95, Azure, BJ Nemeth, BKman74, BMRR, BS 0013, Backtothemacman32, Baderyp, Badgernet, Badjapkid, Baggyeyescar, Barefootguru, Barek, Bart133, Baseballfan, Bawolff, Beano, BenFrantzDale, Bencey, Benjamin Mako Hill, Bento00, Bethx69x, Bholmes6, Big Bird, Bigbluefish, BioHazard417, Bjarkith, Blah2, Blanchardb, Blizzard777, BlueCanary9999, BlueDevil, Bmusician, Bobbo, Bobby Tables, Bobo192, Boccobrock, Boereck, Bonadea, Boogster, Borisblue, Boskoman, Brandon, Brandon J. Marcellus, Bravia2, Brent Wilkins, Brian Crawford, Brianga, Brifendler4, Brittanystein, Brittflick 1994, BrownHornet21, Brownsauce, Brucethemoose, Bsstein, BuickCenturyDriver, BullRangifer, Bunnyhop11, Butterfly0fdoom, Bwrobinett, C.Fred, C628, CIreland, CWii, Cabeman, Calabe1992, Calliegal, Calmer Waters, Calve99, Camac129, Cambookpro, Can't sleep, clown will eat me, CanadianLinuxUser, Canley, Canterbury Tail, Canthusus, Capricorn42, Captain-tucker, CardinalDan, Carlroller, Catgut, Cdc, CesarB, Chabuk, CharlotteWebb, Chazsza, Cheesyperson1995, Chris 73, Chris the speller, ChrisAlsan, Chroniclev, Cipher, Cire100, Cireshoe, Citicat, Ck lostsword, Clawson, ClonedPickle, Cloverboy19, Codingmasters, Cometstyles, CommonsDelinker, Compaq1011, Corti, Crestcala7, Crohnie, Csaag, Curtmack, Cyrus XIII, Czj, D.c.camero, D1536, DAJF, DARTH SIDIOUS 2, DCSkaterkid13, DGG, DalekClock, Dampfnudel, Dancter, Danib62, Daniel Lawrence, Danielign9, Danity94, DaraH46, Darkage7, Darkquota, Darwin-rover, Darwinunicorns, Dasani, Dasher246, David, Davidelit, Davidhorman, Dawnseeker2000, Dcorzine, DeLarge, Dearac, DeathShot39, Decrypt-the-world, Decrypt3, Deepred6502, Dekisugi, Delta Spartan, Deltabeignet, Dennis Bratland, DerHexer, DerekMorr, Dharmabum420, Diligent Terrier, Dindo94, Dirtyterdy, Discospinster, Dishwasherrat, Djmckee1, Djpinoy94, Dlenopy, DocWatson42, Doctoroxenbriery, Dogcutter, Dougw03, Doyler, Dragon76, Drbreznjev, Dreamlogic, Drewray, DropDeadGorgias, Dueyfinster, Duke of ism, Dylan anglada, Dylanspronck, Dysprosia, E147026, EAi, EEMIV, EJF (huggle), ESkog, EVula, EWikist, Ed g2s, EdgeOfEpsilon, Editfinder, Edward301, Eerwitt, Eeshkol, El cangri386, ElKevbo, Eloquence, EnglishEfternamn, Ente75, Enviroboy, Ephilei, Epson291, Eptin, Eran of Arcadia, Eraserhead1, Eric-Wester, Erik9, EryZ, Esanchez7587, Escape Orbit, Essercizi, Euryalus, Evolatable, Evosoho, Exvicious, F1racer101, Falcon8765, Falcon9x5, Family von stennett, Fdny516, Fetchcomms, Fieldday-sunday, Fighting for Justice, Filipão, Fishback, Fivenug, Fletcher, Flewis, Florian Blaschke, Flrn, Fluffernutter, Fluteflute, Fourthords, FrYGuY, Frankenpuppy, Frap, Fraylling, Freakmighty, Fredrick Pendragon, Freegon427, Frencheese, Frijole, Fritz Saalfeld, FritzG, Funke, Future Perfect at Sunrise, Fuzzie, G4rce, GAMBITv5, GDonato, Gabriel, Gamekid, Gateman1997, Gazimoff, Gblaz, Gdo01, Gh5046, Ghent, Gilliam, Glennforever, Globelovers, Gnomon, Godlord2, Gogo Dodo, Goheels681, GoldDragon, Goldenrowley, Gollou, GorillaWarfare, Gorillazx1, Grafen, Graham87, GrahamColm, Grahamstewart, Green Hill, GregorB, Groink, Gyrferret, H.sand01, HJ Mitchell, HPJoker, Habbie, Hahaaa2, Hamish2k, Happinessiseasy, HappyWanderer, Harry, HarryKG, Harryboyles, Haseo9999, Hateless, Hatrox, HawkeAnyone, Hayden120, Hdt83, Hellisp, Hello32020, Helzagood, Henry Flower, HereToHelp, Herr Anonymus, Hghyux, HindenPeter, Hirohisat, Hooperbloob, Horserice, Htmnssn, Huangcjz, Husond, Hut 8.5, Huwr, Hydrogen Iodide, Hyperchiken, I-Podfreak123, IE, IPave, IW.HG, Iain99, Iamdigitalman, Iceberg3k, Igoldste, Iheartrobojesus, Ilikecats1234, Ilikefood, Indon, Interchangeable, Iphonefans2009, Ipodnanomart, Ipodsrule, Ipodsucks, Iridescent, Irregulargalaxies, Irvine.kwong, Isaidgoodday, It Is Me Here, Iuhkjhk87y678, Ixfd64, Izno, J Di, J.delanoy, JACK8411, JFM11613, JForget, JLaTondre, JMartin46, JQF, JRomero, JV Smithy, Jack007, Jackson22, Jacoplane, JaffaCakeLover, Jalabi99, JamesEncliffe, Jameshfisher, Jameswa21, Jasca Ducato, Jason12345, Java13690, Jcam962, Jcw69, Jdot182, Jedi94, Jedibob5, Jeff G., Jennavecia, Jeremy Banks, Jesant13, Jj137, Jjnguy, Jjupiter100, Jkelly, Jl1995, Jmclark911, Joelvanatta, JohJak2, JohnnyBGood, Jojhutton, JonHarder, JonT, Jonathan Webley, Jons63, Joz3d, Jp3z, Jsquigs21, Jtalledo, Jtbean, Jumping cheese, JunKami, Just James, Justaperson117, Jw12321, Jw21, Jyounguxd, K. Annoyomous, KDesk, KGasso, Kamasutra, Karl pierson, Keakealani, Keegscee, Keilana, KeithB, KelleyCook, Kennethmac2000, Kerotan, Kharabgulab, Khrister, Killa-Kay16, Killipod, Kilo-Lima, Kinaro, King Cole, Kinimv, Kitch, Klubclub1, Kniveswood, KnowledgeOfSelf, Koolness10, Kpr00, Kristy96, Kruhw, KsbjA, Ksdhindsa, Kungming2, KyleFTW, Kyro, LAX, Lakers, Lar, Latitude0116, LeaveSleaves, Legend78, Leo-Roy, Leuko, Lex839, Lil Lez, Limetolime, Little Mountain 5, Littledoggy26, Littlemisscookie, Lonaowna, Longhair, Lordofchaosiori, Louis Do Nothing, LovesMacs, Lppa, Lt. penguin, Lubyanka, Luk, Luna Santin, M.O.X, M1ss1ontomars2k4, MER-C, Macaddict10, MacintoshWriter, Madchester, Madhero88, Maf5081, Mahanga, Makatota, Malcolmxl5, Mandarax, MarcoA.G.B, Marek69, Mariacer Cervantes, Markt3, Marrsbty, Martarius, Mary Hope, Masterhand10, Materialscientist, Maths & Science legend, Matt Yohe, Matthew98, MatthewWaller, MattieTK, Mattk39, Mattop123, Mavericklovesipod, Maxis ftw, Mayor Coffee Bean, Mclay1, Mcmatter, Mdzarate, Meaghan, Meateater, Meeples, Megamanfan3, Melsaran, Mentifisto, Mermaid from the Baltic Sea, Michaelmuffin, Michaelzeng7, Mickeys1fan, Mifter, Mikethepom, Minorkeys88, Mintchocicecream, Misbeliever, Mitchello, Mitsuki Mirashima Hyati, Mmdoogie, Moe Epsilon, Mono, Moreno Valley User, Moronicles, Mossj002, Mp3hist, Mpcpaul, Mpk, Mr.Verwalter, Mr.bombo, MrFish, Mrpmpfan, Mrwojo, Muhandes, Mushroom, Muéro, Mwarren us, Mynameinc, Myncknm, N i grunze, NXTguru, Nakomaru, Nanami Kamimura, Nandesuka, Nanoguy85, Nathaniel, NatureA16, NawlinWiki, NeoNerd, NerdyScienceDude, NewYorkDreams, Nicofulto2010, Nihiltres, Nintendude, Niro The Crazy, Nneonneo, Nodekeeper, Nolan.carrozza, NotsraM, Nrbelex, Nuiop729, Nuttycoconut, O.neill.kid, OSborn, Obafgkm, Obli, Of, Ohconfucius, Omegamormegil, Omghihowareyou, Onorem, Orii, Oscarthecat, Otzithecaveman, Oxymoron83, Ozphactor, P.B. Pilhet, PDH, PM800, Papre, Pauerbach08, Pboy2k5, Peanut4, Pekaje, Peruvianllama, Peterrobbemond, Peterson8765, Peteturtle, Petri Krohn, Pfhunt, Pgk, Phaneuf3, Phaser2, PhilKnight, Philip Trueman, Phmall, Photographerguy, Piano non troppo, Pichu1988, PoiZaN, Polarman..AaA, Poopoo12332112345678910, Poopypants89475, Possum, Pretzels, ProductFixya, ProductResource, Professorlui, ProhibitOnions, Psantora, Psykeeper, Puddleglum42, Pundit, Pyrocrickett, Q11z, Qtac, Quadratus, Quake max, Quasipalm, Quau, Qweqazbob, Qwerty Binary, Qxz, Racefreak265, RadicalBender, RadioActive, Raekwon, Rahizzy, Rama's Arrow, Ramin987, Randomwellwisher, Raptorjesuspwns, Rattlesnake, Redds4ever, Redrunner89, Rehas, Rehevkor, Reil, Renatta81, Res2216firestar, RetiredUser124642196, Revrit, Riana, Rich Farmbrough, Richard Wolf VI, Rikuansem13, Ripcurl-lover, Rjd0060, Rjwilmsi, Roastytoast, Rob Cranfill, Robert A. Mitchell, Roberthunt, Robertvan1, Rockymountains, Rockysmile11, RodC, Rogerd, Romanian Relish, Ronhjones, Rorschach, Rror, Rufous, Rush2009, RxS, Ryu Ematsu, SA Foxxe, SMC, ST47, Sagaciousuk, Sahaskatta, SaiyanEmperor2008, SaltyBoatr, Sam Clark, Sam916, SamCostello, Samadam, Samdutton, Samir, Samwb123, Sango123, Santa.on.Wiki, Sardanaphalus, Sarken, Sarranduin, Sarrus, Saxsux, Scepia, Scetoaux, Schaefer, SchfiftyThree, Schomorox, SchuminWeb, Scootey, ScottSteiner, Screech123, Sdfisher, Seamonkey237, Sean William, Seaphoto, Sechzehn, Sedidome, Sesu Prime, Sevua, Seweri, Sgs101, Shalroth, Sharadbob, ShaunL, Shawnc, Sheeana, Shell Kinney, Shilly42, Shiori, Shlomke, Shocut, Shshshsh, Sigma 7, SignalMan, Sintaku, Sir dorkus, Siryendor, Skew-t, Skier Dude, Skremer, Skunkape, Skybon, Skyeap, Slick08, Slowking Man, Snausageszzzzzzz, Snowolf, Sonicblue4, Sonyps2, Sooperfalcon8, Sosekopp, Soundslikealotofhooplah, Speer320, Spesh531, Spinecraft, Splintax, Spongesquid, Steel, Stepa, Steven X, Stevencho, Stevenguerin, Steverapaport, Stroppolo, Supasnake, Superman 1234567, Swiftfoxlit, TJ Donlan, TJ Spyke, Tad Lincoln, Takaitra, Tamajared, Tanner 183, Tapir666, Targowski, Tariqabjotu, Tcardone05, Tcncv, Techey123, Teles, Teorik Deli, Thaurisil, The Anome, The Evil Spartan, The Giant Puffin, The Ronin, The Thing That Should Not Be, The Utahraptor, The beef, The slick mick, The undertow, The33dude33, The729, TheGreatConspiracy, TheIguana, Theaveng, Thegreenj, Theunixgeek, Thief12, Thingg, Thmars10, Thoughtfix, Thrane, Thrashmeister, Thumperward, Tiddly Tom, Tide rolls, Timandrews, Tktktk, ToNToNi, Tobi paull, Tokek, Tommy2010, Tomnason1010, Towel401,

Trailer4who, Trendsettr, Tripsplant, Truthmaster0, Tyrantstuff, Ucanlookitup, Ulric1313, Urgedgyro, Urhixidur, Usb man, Useight, Utkarsh apple, Vadde, VanillaDeath, Vegpuff, Versus22, Vickypathak1990, Vincent Pun, Vinibinini, Vinnivince, ViperSnake151, VisvambaNathan, Vladsinger, Wackymacs, Wanna Know My Name? Later, Wannger27, Watch The Beef, Wayne Slam, Wcrofus, WebHamster, Wendell, Wheelboss, White 720, WhiteCat, Why Not A Duck, Wikifan21century, Wikimanisawesome, Wikisplasher, Willking1979, Winchelsea, Wknight94, WoodenFeet, Woodshed, Woohookitty, Wrh2, Wrockca, Ww.ellis, Wwwilf wilson, Xeno, Xevious, Xiahou, Xp54321, Yamaguchi, Yamfox, Yeah0011, Yellowdesk, Yelyos, Yimeng, Yinyang1195, Yoahann, YuMaNuMa, ZOMGravity, Zanndostaro, Zeno333, Zenohockey, Zetawoof, Zhelfinstein, Zidane121, Ziggaroth, Zimshallrule, Zj morales, Zodon, Zonafan39, Zundark, Zxcvasdfqwerbgtnhymju, Zxcvboy97, 2165 anonymous edits

# Image Sources, Licenses and Contributors

**Image:SendStation logo.png** *Source*: http://en.wikipedia.org/w/index.php?title=File:SendStation_logo.png *License*: unknown *Contributors*: User:Cydebot, User:Dr. Woo

**File:iPod family.png** *Source*: http://en.wikipedia.org/w/index.php?title=File:IPod_family.png *License*: unknown *Contributors*: Kyro

**File:Various iPods.jpg** *Source*: http://en.wikipedia.org/w/index.php?title=File:Various_iPods.jpg *License*: unknown *Contributors*: Chris Harrison from Augusta, GA, USA

**File:IPod wheel.svg** *Source*: http://en.wikipedia.org/w/index.php?title=File:IPod_wheel.svg *License*: unknown *Contributors*: Original uploader was Googleaseerch at en.wikipedia Later version(s) were uploaded by CountingPine at en.wikipedia.

**File:4 iPod Chargers.JPG** *Source*: http://en.wikipedia.org/w/index.php?title=File:4_iPod_Chargers.JPG *License*: unknown *Contributors*: User:HereToHelp

**File:Made for iPod.svg** *Source*: http://en.wikipedia.org/w/index.php?title=File:Made_for_iPod.svg *License*: unknown *Contributors*: Apple Inc. (original uploader was Koman90 at en.wikipedia)

**File:IPod Earbuds.JPG** *Source*: http://en.wikipedia.org/w/index.php?title=File:IPod_Earbuds.JPG *License*: unknown *Contributors*: Epson291, HereToHelp, MB-one, Shoulder-synth, 1 anonymous edits

**File:ipod 1G.png** *Source*: http://en.wikipedia.org/w/index.php?title=File:Ipod_1G.png *License*: unknown *Contributors*: Original uploader was Rjcflyer@aol.com at en.wikipedia

**File:ipod 2G.png** *Source*: http://en.wikipedia.org/w/index.php?title=File:Ipod_2G.png *License*: unknown *Contributors*: Original uploader was Rjcflyer@aol.com at en.wikipedia

**File:ipod backlight transparent.png** *Source*: http://en.wikipedia.org/w/index.php?title=File:Ipod_backlight_transparent.png *License*: unknown *Contributors*: User:Ilmari Karonen, User:Xell

**File:iPod4G.jpg** *Source*: http://en.wikipedia.org/w/index.php?title=File:IPod4G.jpg *License*: unknown *Contributors*: Original uploader was Lamename3000 at en.wikipedia

**File:iPodphoto4G 1.png** *Source*: http://en.wikipedia.org/w/index.php?title=File:IPodphoto4G_1.png *License*: unknown *Contributors*: Original Photograph - AquaStreakImage Cleanup - Rugby471

**File:ipod 5th Generation white rotated.png** *Source*: http://en.wikipedia.org/w/index.php?title=File:Ipod_5th_Generation_white_rotated.png *License*: unknown *Contributors*: User:Ilmari Karonen

**File:iPod classic.png** *Source*: http://en.wikipedia.org/w/index.php?title=File:IPod_classic.png *License*: unknown *Contributors*: User:Kyro

**File:Green ipodmini 1stgen.jpg** *Source*: http://en.wikipedia.org/w/index.php?title=File:Green_ipodmini_1stgen.jpg *License*: unknown *Contributors*: JayKeaton, Vipersnake151, 4 anonymous edits

**File:Green ipodmini 2ndgen.jpg** *Source*: http://en.wikipedia.org/w/index.php?title=File:Green_ipodmini_2ndgen.jpg *License*: unknown *Contributors*: FlickreviewR, JayKeaton, Riana, Túrelio, 5 anonymous edits

**File:iPod Nano in its Dock.png** *Source*: http://en.wikipedia.org/w/index.php?title=File:IPod_Nano_in_its_Dock.png *License*: unknown *Contributors*: User:Keeleysam

**File:Blue iPod Nano.jpg** *Source*: http://en.wikipedia.org/w/index.php?title=File:Blue_iPod_Nano.jpg *License*: unknown *Contributors*: HereToHelp

**File:iPod nano 3g black.jpg** *Source*: http://en.wikipedia.org/w/index.php?title=File:IPod_nano_3g_black.jpg *License*: unknown *Contributors*: Andrew from London, UK; modified by

**File:iPod Nano 4G black crop.jpg** *Source*: http://en.wikipedia.org/w/index.php?title=File:IPod_Nano_4G_black_crop.jpg *License*: unknown *Contributors*: User:Aconcagua

**File:IPod-Nano-5G-front-back.png** *Source*: http://en.wikipedia.org/w/index.php?title=File:IPod-Nano-5G-front-back.png *License*: unknown *Contributors*: Lubyanka

**File:6G IPod Nano.svg** *Source*: http://en.wikipedia.org/w/index.php?title=File:6G_IPod_Nano.svg *License*: unknown *Contributors*: User:House

**File:iPod shuffle 1G.png** *Source*: http://en.wikipedia.org/w/index.php?title=File:IPod_shuffle_1G.png *License*: unknown *Contributors*: User:Kyro

**File:iPod Shuffle Crop.jpg** *Source*: http://en.wikipedia.org/w/index.php?title=File:IPod_Shuffle_Crop.jpg *License*: unknown *Contributors*: HereToHelp

**File:iPod shuffle 3G.png** *Source*: http://en.wikipedia.org/w/index.php?title=File:IPod_shuffle_3G.png *License*: unknown *Contributors*: User:Kyro

**File:5G IPod Shuffle.svg** *Source*: http://en.wikipedia.org/w/index.php?title=File:5G_IPod_Shuffle.svg *License*: unknown *Contributors*: User:House

**File:Ipod-touch-1st-gen.jpg** *Source*: http://en.wikipedia.org/w/index.php?title=File:Ipod-touch-1st-gen.jpg *License*: unknown *Contributors*: User:Evan-Amos

**File:IPod_touch_2G.png** *Source*: http://en.wikipedia.org/w/index.php?title=File:IPod_touch_2G.png *License*: unknown *Contributors*: User:Kyro

**File:IPod_touch_4G.png** *Source*: http://en.wikipedia.org/w/index.php?title=File:IPod_touch_4G.png *License*: unknown *Contributors*: Matthieu Riegler

**File:Ipod sales per quarter.svg** *Source*: http://en.wikipedia.org/w/index.php?title=File:Ipod_sales_per_quarter.svg *License*: unknown *Contributors*: User:MySchizoBuddy

**File:Macintosh 128k transparency.png** *Source*: http://en.wikipedia.org/w/index.php?title=File:Macintosh_128k_transparency.png *License*: unknown *Contributors*: Admrboltz, Avian, Brossow, Common Good, Grm wnr, Herbythyme, Himasaram, Kenmayer, Kevin chen2003, Mono, Pharos, 4 anonymous edits

**File:Imac 16-9.png** *Source*: http://en.wikipedia.org/w/index.php?title=File:Imac_16-9.png *License*: unknown *Contributors*: + Background : , under CC-by 3.0

**File:Loudspeaker.svg** *Source*: http://en.wikipedia.org/w/index.php?title=File:Loudspeaker.svg *License*: unknown *Contributors*: Bayo, Gmaxwell, Husky, Iamunknown, Mirithing, Myself488, Nethac DIU, Omegatron, Rocket000, The Evil IP address, Wouterhagens, 18 anonymous edits

**Image:Apple Macintosh Desktop.png** *Source*: http://en.wikipedia.org/w/index.php?title=File:Apple_Macintosh_Desktop.png *License*: unknown *Contributors*: A104375, AlistairMcMillan, BorgQueen, Chmod007, Ctachme, Diego Moya, EdC, Ferrenrock, Grm wnr, HereToHelp, N. Harmonik, Ricardo Cancho Niemietz, 3 anonymous edits

**File:Ad apple 1984.jpg** *Source*: http://en.wikipedia.org/w/index.php?title=File:Ad_apple_1984.jpg *License*: unknown *Contributors*: Classicfilms, Edderso, Geniac, Grm wnr, HereToHelp, Ilse@, Jamoche, Lars T., Melesse, Mendaliv, Psantora, Rimshot, Zanimum, 5 anonymous edits

**Image:Apple-Macintosh.jpg** *Source*: http://en.wikipedia.org/w/index.php?title=File:Apple-Macintosh.jpg *License*: unknown *Contributors*: User:Tmarki

**Image:MacII.jpg** *Source*: http://en.wikipedia.org/w/index.php?title=File:MacII.jpg *License*: unknown *Contributors*: Grm wnr, Joonasl

**Image:Macintosh portable.jpg** *Source*: http://en.wikipedia.org/w/index.php?title=File:Macintosh_portable.jpg *License*: unknown *Contributors*: Beavis, Grm wnr, J o, Kozuch, Kyro, Ranveig, Wpedzich, Wutsje, 1 anonymous edits

**Image:Macintosh classic.jpg** *Source*: http://en.wikipedia.org/w/index.php?title=File:Macintosh_classic.jpg *License*: unknown *Contributors*: Grm wnr, Kilom691, Tipiac

**Image:Powerbook 100 pose.jpg** *Source*: http://en.wikipedia.org/w/index.php?title=File:Powerbook_100_pose.jpg *License*: unknown *Contributors*: User:Danamania

**Image:Macintosh System 7.5.3 screenshot.png** *Source*: http://en.wikipedia.org/w/index.php?title=File:Macintosh_System_7.5.3_screenshot.png *License*: unknown *Contributors*: Diego Moya, McLoaf, N. Harmonik, Remember the dot, 1 anonymous edits

**File:MacOS8.6.jpg** *Source*: http://en.wikipedia.org/w/index.php?title=File:MacOS8.6.jpg *License*: unknown *Contributors*: Calmer Waters, Fetchcomms, Fourthords, HereToHelp, Mono, N. Harmonik, Ozguy89

**Image:IMac Bondi Blue.jpg** *Source*: http://en.wikipedia.org/w/index.php?title=File:IMac_Bondi_Blue.jpg *License*: unknown *Contributors*: Apalsola, Dodo, Grm wnr, HereToHelp, Ilse@, Thuresson

**Image:MacBook Pro situated on a wooden table.jpg** *Source*: http://en.wikipedia.org/w/index.php?title=File:MacBook_Pro_situated_on_a_wooden_table.jpg *License*: unknown *Contributors*: Jeff Geerling

**File:Mac mini.png** *Source*: http://en.wikipedia.org/w/index.php?title=File:Mac_mini.png *License*: unknown *Contributors*: User:Wiki637

**File:Macpro BW.png** *Source*: http://en.wikipedia.org/w/index.php?title=File:Macpro_BW.png *License*: unknown *Contributors*: Bayo, Bukk, Kyro, Max Naylor

**File:MacBook Air.png** *Source*: http://en.wikipedia.org/w/index.php?title=File:MacBook_Air.png *License*: unknown *Contributors*:

**Image:MacBook Pros.jpg** *Source*: http://en.wikipedia.org/w/index.php?title=File:MacBook_Pros.jpg *License*: unknown *Contributors*: Benjamin Nagel

**Image:IMacG5guts.png** *Source*: http://en.wikipedia.org/w/index.php?title=File:IMacG5guts.png *License*: unknown *Contributors*: Grm wnr

**File:Audi logo detail.svg** *Source*: http://en.wikipedia.org/w/index.php?title=File:Audi_logo_detail.svg *License*: unknown *Contributors*: Dmitry89, Mahdipur, Ysangkok, Егор Осин, 1 anonymous edits

**File:Increase2.svg** *Source*: http://en.wikipedia.org/w/index.php?title=File:Increase2.svg *License*: unknown *Contributors*: Sarang

**File:Audi Typ E (1923).jpg** *Source*: http://en.wikipedia.org/w/index.php?title=File:Audi_Typ_E_(1923).jpg *License*: unknown *Contributors*: Alexander Z., Infrogmation, MartinHansV, Mattes, Olivier2

**File:IFA F9 front 20040924.jpg** *Source*: http://en.wikipedia.org/w/index.php?title=File:IFA_F9_front_20040924.jpg *License*: unknown *Contributors*: user:Randy43

**File:Bundesarchiv B 145 Bild-F038791-0035, Wolfsburg, VW Autowerk.jpg** *Source*: http://en.wikipedia.org/w/index.php?title=File:Bundesarchiv_B_145_Bild-F038791-0035,_Wolfsburg,_VW_Autowerk.jpg *License*: unknown *Contributors*: Schaack, Lothar

**File:Audi quattro.jpg** *Source*: http://en.wikipedia.org/w/index.php?title=File:Audi_quattro.jpg *License*: unknown *Contributors*: Anetode, MB-one, Wikisearcher

**File:Audi 100 C3 front 20080212.jpg** *Source*: http://en.wikipedia.org/w/index.php?title=File:Audi_100_C3_front_20080212.jpg *License*: unknown *Contributors*: user:Randy43

**File:Audi aurangabad.jpg** *Source*: http://en.wikipedia.org/w/index.php?title=File:Audi_aurangabad.jpg *License*: unknown *Contributors*: DIGIADS

**File:Audi R8 front.JPG** *Source*: http://en.wikipedia.org/w/index.php?title=File:Audi_R8_front.JPG *License*: unknown *Contributors*: Taco325i

**Image:Volkswagen W12.jpg** *Source*: http://en.wikipedia.org/w/index.php?title=File:Volkswagen_W12.jpg *License*: unknown *Contributors*: User Hasse A on sv.wikipedia

**File:2007 Audi R8 Engine.JPG** *Source*: http://en.wikipedia.org/w/index.php?title=File:2007_Audi_R8_Engine.JPG *License*: unknown *Contributors*: User:Hatsukari715

**File:LED DaytimeRunningLights.jpg** *Source*: http://en.wikipedia.org/w/index.php?title=File:LED_DaytimeRunningLights.jpg *License*: unknown *Contributors*: User:Thermos

**File:Audi A1 1.4 TFSI Ambition S-line – Frontansicht, 15. Mai 2011, Wuppertal.jpg** *Source*: http://en.wikipedia.org/w/index.php?title=File:Audi_A1_1.4_TFSI_Ambition_S-line_–_Frontansicht,_15._Mai_2011,_Wuppertal.jpg *License*: unknown *Contributors*: user:M 93

**File:Audi A3 Sportback front 20090714.jpg** *Source*: http://en.wikipedia.org/w/index.php?title=File:Audi_A3_Sportback_front_20090714.jpg *License*: unknown *Contributors*: user:Randy43

**File:Audi A4 sedan -- 02-18-2011.jpg** *Source*: http://en.wikipedia.org/w/index.php?title=File:Audi_A4_sedan_--_02-18-2011.jpg *License*: unknown *Contributors*: IFCAR

**File:2010 Audi A5 (8T) 3.0 TDI quattro Sportback 03.jpg** *Source*: http://en.wikipedia.org/w/index.php?title=File:2010_Audi_A5_(8T)_3.0_TDI_quattro_Sportback_03.jpg *License*: unknown *Contributors*: User:OSX

**File:2012 Audi A6 -- 07-07-2011 2.jpg** *Source*: http://en.wikipedia.org/w/index.php?title=File:2012_Audi_A6_--_07-07-2011_2.jpg *License*: unknown *Contributors*: IFCAR

**File:2012 Audi A7 -- 07-07-2011 1.jpg** *Source*: http://en.wikipedia.org/w/index.php?title=File:2012_Audi_A7_--_07-07-2011_1.jpg *License*: unknown *Contributors*: IFCAR

**File:Audi A8 D4 registered September 2010 4134 cc Diesel 4wd.jpg** *Source*: http://en.wikipedia.org/w/index.php?title=File:Audi_A8_D4_registered_September_2010_4134_cc_Diesel_4wd.jpg *License*: unknown *Contributors*: User:Charles01

**File:2007 Audi TT (8J) 2.0 TFSI coupe 01.jpg** *Source*: http://en.wikipedia.org/w/index.php?title=File:2007_Audi_TT_(8J)_2.0_TFSI_coupe_01.jpg *License*: unknown *Contributors*: User:OSX

**File:Audi R8 V10 – Frontansicht, 13. März 2011, Wuppertal.jpg** *Source*: http://en.wikipedia.org/w/index.php?title=File:Audi_R8_V10_–_Frontansicht,_13._März_2011,_Wuppertal.jpg *License*: unknown *Contributors*: user:M 93

**File:Audi Q3 2.0 TDI quattro S tronic Samoaorange.JPG** *Source*: http://en.wikipedia.org/w/index.php?title=File:Audi_Q3_2.0_TDI_quattro_S_tronic_Samoaorange.JPG *License*: unknown *Contributors*: User:Thomas doerfer

**File:Audi Q5 front 20090404.jpg** *Source*: http://en.wikipedia.org/w/index.php?title=File:Audi_Q5_front_20090404.jpg *License*: unknown *Contributors*: user:Randy43

**File:Audi Q7 (Facelift) front 20110115.jpg** *Source*: http://en.wikipedia.org/w/index.php?title=File:Audi_Q7_(Facelift)_front_20110115.jpg *License*: unknown *Contributors*: user:M 93

**File:Audi S3 Sprintblau.JPG** *Source*: http://en.wikipedia.org/w/index.php?title=File:Audi_S3_Sprintblau.JPG *License*: unknown *Contributors*: User:Thomas doerfer

**File:Audi S4 B8 Kombi front 20090412.jpg** *Source*: http://en.wikipedia.org/w/index.php?title=File:Audi_S4_B8_Kombi_front_20090412.jpg *License*: unknown *Contributors*: user:Randy43

**File:AudiS five.jpg** *Source*: http://en.wikipedia.org/w/index.php?title=File:AudiS_five.jpg *License*: unknown *Contributors*: Kalle Anka (Michael Meinecke)

**File:Audi S6 C6 front 2010 0514.jpg** *Source*: http://en.wikipedia.org/w/index.php?title=File:Audi_S6_C6_front_2010_0514.jpg *License*: unknown *Contributors*: user:Randy43

**File:Audi TT S -- 2011 DC.jpg** *Source*: http://en.wikipedia.org/w/index.php?title=File:Audi_TT_S_--_2011_DC.jpg *License*: unknown *Contributors*: IFCAR

**File:Audi RS3 Sportback 2.5 TFSI quattro S tronic Misanorot.JPG** *Source*: http://en.wikipedia.org/w/index.php?title=File:Audi_RS3_Sportback_2.5_TFSI_quattro_S_tronic_Misanorot.JPG *License*: unknown *Contributors*: User:Thomas doerfer

**File:Audi-RS4.jpg** *Source*: http://en.wikipedia.org/w/index.php?title=File:Audi-RS4.jpg *License*: unknown *Contributors*: User:Samoborac

**File:Audi RS5 front.jpg** *Source*: http://en.wikipedia.org/w/index.php?title=File:Audi_RS5_front.jpg *License*: unknown *Contributors*: Molch-Entertainment at de.wikipedia

**File:Audi TT-RS.jpg** *Source*: http://en.wikipedia.org/w/index.php?title=File:Audi_TT-RS.jpg *License*: unknown *Contributors*: User:Ascaron

**File:Portugal 84 Audi Quattro A2.jpg** *Source*: http://en.wikipedia.org/w/index.php?title=File:Portugal_84_Audi_Quattro_A2.jpg *License*: unknown *Contributors*: User:MPW57

**File:Audi Quattro - 2007 Rallye Deutschland (cropped).jpg** *Source*: http://en.wikipedia.org/w/index.php?title=File:Audi_Quattro_-_2007_Rallye_Deutschland_(cropped).jpg *License*: unknown *Contributors*: realname

**File:Audi R10 1.JPG** *Source*: http://en.wikipedia.org/w/index.php?title=File:Audi_R10_1.JPG *License*: unknown *Contributors*: Fabrice Pluchet

**File:Audi Logo.svg** *Source*: http://en.wikipedia.org/w/index.php?title=File:Audi_Logo.svg *License*: unknown *Contributors*: Avicennasis, Jovianeye

**File:Audi Centre Sydney, Zetland, New South Wales (2010-07-10) 02.jpg** *Source*: http://en.wikipedia.org/w/index.php?title=File:Audi_Centre_Sydney,_Zetland,_New_South_Wales_(2010-07-10)_02.jpg *License*: unknown *Contributors*: User:OSX

**File:Audi R8 3 Quarter.jpg** *Source*: http://en.wikipedia.org/w/index.php?title=File:Audi_R8_3_Quarter.jpg *License*: unknown *Contributors*: User:Niyathy

**File:Wrigley Logo.svg** *Source*: http://en.wikipedia.org/w/index.php?title=File:Wrigley_Logo.svg *License*: unknown *Contributors*: unknown

**Image:Wrigley Building - Chicago, Illinois.JPG** *Source*: http://en.wikipedia.org/w/index.php?title=File:Wrigley_Building_-_Chicago,_Illinois.JPG *License*: unknown *Contributors*: Daderot

**File:Wrigley's Gum 1920 Ad.jpg** *Source*: http://en.wikipedia.org/w/index.php?title=File:Wrigley's_Gum_1920_Ad.jpg *License*: unknown *Contributors*: Wrigley's

**Image:Personal computer, exploded 5.svg** *Source*: http://en.wikipedia.org/w/index.php?title=File:Personal_computer,_exploded_5.svg *License*: unknown *Contributors*: Amada44, BRUTE, Gengiskanhg, Gustavb, Herbythyme, Hidro, Homonihilis, Jikayo, Kozuch, Mdd, Mhby87, Pepy, Petr.wudi, PlusMinus, Shmecherul70, Túrelio, خالد حسنی, 34 anonymous edits

**File:Computer from inside 018.jpg** *Source*: http://en.wikipedia.org/w/index.php?title=File:Computer_from_inside_018.jpg *License*: unknown *Contributors*: User:Kallerna

**Image:3-Tastenmaus Microsoft.jpg** *Source*: http://en.wikipedia.org/w/index.php?title=File:3-Tastenmaus_Microsoft.jpg *License*: unknown *Contributors*: Aka, Darkone, GreyCat, Warden

**Image:Apple Dock Connector.jpg** *Source*: http://en.wikipedia.org/w/index.php?title=File:Apple_Dock_Connector.jpg *License*: unknown *Contributors*: User: Synthesis Studios on Flickr.

**Image:IPodclassiclogo.png** *Source*: http://en.wikipedia.org/w/index.php?title=File:IPodclassiclogo.png *License*: unknown *Contributors*: Apple Inc.

**File:6G iPod.svg** *Source*: http://en.wikipedia.org/w/index.php?title=File:6G_iPod.svg *License*: unknown *Contributors*: Eduloqui, Kyro, Rocket000

**Image:Lightmatter ipod 1G.jpg** *Source*: http://en.wikipedia.org/w/index.php?title=File:Lightmatter_ipod_1G.jpg *License*: unknown *Contributors*: Aaron Logan

**Image:Ipod 2G.png** *Source*: http://en.wikipedia.org/w/index.php?title=File:Ipod_2G.png *License*: unknown *Contributors*: Original uploader was Rjcflyer@aol.com at en.wikipedia

**Image:3G ipod in dock.jpg** *Source*: http://en.wikipedia.org/w/index.php?title=File:3G_ipod_in_dock.jpg *License*: unknown *Contributors*: Ed g2s, Epson291, Kenmayer, Luki.r

**Image:Ipod4g.JPG** *Source*: http://en.wikipedia.org/w/index.php?title=File:Ipod4g.JPG *License*: unknown *Contributors*: User Loft on en.wikipedia

**Image:Ipod 5th Generation white.jpg** *Source*: http://en.wikipedia.org/w/index.php?title=File:Ipod_5th_Generation_white.jpg *License*: unknown *Contributors*: User:Stahlkocher

**File:20070913 iPod 5-6 Gen side-by-side.jpg** *Source*: http://en.wikipedia.org/w/index.php?title=File:20070913_iPod_5-6_Gen_side-by-side.jpg *License*: unknown *Contributors*: User:TonyTheTiger. Original uploader was TonyTheTiger at en.wikipedia

**Image:ipodclassic120gb.jpg** *Source*: http://en.wikipedia.org/w/index.php?title=File:Ipodclassic120gb.jpg *License*: unknown *Contributors*: e29616

**Image:Ipod 1G.png** *Source*: http://en.wikipedia.org/w/index.php?title=File:Ipod_1G.png *License*: unknown *Contributors*: Original uploader was Rjcflyer@aol.com at en.wikipedia

**Image:Ipod backlight transparent.png** *Source*: http://en.wikipedia.org/w/index.php?title=File:Ipod_backlight_transparent.png *License*: unknown *Contributors*: User:Ilmari Karonen, User:Xell

**Image:IPodphoto4G 1.png** *Source*: http://en.wikipedia.org/w/index.php?title=File:IPodphoto4G_1.png *License*: unknown *Contributors*: Original Photograph - AquaStreakImage Cleanup - Rugby471

**Image:iPodphoto4G 1.png** *Source*: http://en.wikipedia.org/w/index.php?title=File:IPodphoto4G_1.png *License*: unknown *Contributors*: Original Photograph - AquaStreakImage Cleanup - Rugby471

**File:IPod classic.png** *Source*: http://en.wikipedia.org/w/index.php?title=File:IPod_classic.png *License*: unknown *Contributors*: User:Kyro

**File:IPodtouchlogo.png** *Source*: http://en.wikipedia.org/w/index.php?title=File:IPodtouchlogo.png *License*: unknown *Contributors*: Apple Inc.

**File:IPod touch 4G.png** *Source*: http://en.wikipedia.org/w/index.php?title=File:IPod_touch_4G.png *License*: unknown *Contributors*: Matthieu Riegler

**File:IPod Touch 2G Remote Mic.jpg** *Source*: http://en.wikipedia.org/w/index.php?title=File:IPod_Touch_2G_Remote_Mic.jpg *License*: unknown *Contributors*: User:Fletcher6

**File:Ipod-touch-1st-gen.png** *Source*: http://en.wikipedia.org/w/index.php?title=File:Ipod-touch-1st-gen.png *License*: unknown *Contributors*: User:Beao, User:Evan-Amos

**File:IPod touch 2G.png** *Source*: http://en.wikipedia.org/w/index.php?title=File:IPod_touch_2G.png *License*: unknown *Contributors*: User:Kyro

**File:IPodlogo.png** *Source*: http://en.wikipedia.org/w/index.php?title=File:IPodlogo.png *License*: unknown *Contributors*: -

**File:IPod nano 6G.png** *Source*: http://en.wikipedia.org/w/index.php?title=File:IPod_nano_6G.png *License*: unknown *Contributors*: Kyro

**File:Ipod nano box.jpg** *Source*: http://en.wikipedia.org/w/index.php?title=File:Ipod_nano_box.jpg *License*: unknown *Contributors*: Dodo, Husky, Platonides, Sam916, 1 anonymous edits

**File:IPod-Nano-package.jpg** *Source*: http://en.wikipedia.org/w/index.php?title=File:IPod-Nano-package.jpg *License*: unknown *Contributors*: Ed g2s, FlickreviewR, HereToHelp, Iamunknown

**File:KoolgiyBlackNano.JPG** *Source*: http://en.wikipedia.org/w/index.php?title=File:KoolgiyBlackNano.JPG *License*: unknown *Contributors*: User:Koolgiy

**File:Ipod nano g3 003.jpg** *Source*: http://en.wikipedia.org/w/index.php?title=File:Ipod_nano_g3_003.jpg *License*: unknown *Contributors*: Original uploader was The slick mick at en.wikipedia

**File:IPod Nano 4G black crop.jpg** *Source*: http://en.wikipedia.org/w/index.php?title=File:IPod_Nano_4G_black_crop.jpg *License*: unknown *Contributors*: User:Aconcagua

**File:IPod-nano-5G-camera-view-1.png** *Source*: http://en.wikipedia.org/w/index.php?title=File:IPod-nano-5G-camera-view-1.png *License*: unknown *Contributors*: Lubyanka

**File:1G Nano iPod.svg** *Source*: http://en.wikipedia.org/w/index.php?title=File:1G_Nano_iPod.svg *License*: unknown *Contributors*: Eduloqui, House, Kyro, Rocket000

**File:Ipod nano.JPG** *Source*: http://en.wikipedia.org/w/index.php?title=File:Ipod_nano.JPG *License*: unknown *Contributors*: Original uploader was ShaunL at en.wikipedia

**File:IPod-Nano-5G-front.png** *Source*: http://en.wikipedia.org/w/index.php?title=File:IPod-Nano-5G-front.png *License*: unknown *Contributors*: Lubyanka

MIX
Papier aus verantwortungsvollen Quellen
Paper from responsible sources
FSC® C105338

Printed by Books on Demand GmbH, Norderstedt / Germany